IAASTD

International Assessment of Agricultural Knowledge, Science and Technology for Development

Synthesis Report

IAASTD

International Assessment of Agricultural Knowledge, Science and Technology for Development

UNEP

UNESCO

THE WORLD BANK

WHO

GLOBAL ENVIRONMENT FACILITY

IAASTD

International Assessment of Agricultural Knowledge, Science and Technology for Development

Synthesis Report

A Synthesis of the Global and Sub-Global IAASTD Reports

Edited by

| Beverly D. McIntyre | Hans R. Herren | Judi Wakhungu | Robert T. Watson |
| IAASTD Secretariat | Millennium Institute | African Centre for Technology Studies | University of East Anglia |

Island Press is a trademark of The Center for Resource Economics.

Library of Congress Cataloging-in-Publication data.

International assessment of agricultural knowledge, science and technology for development (IAASTD) : synthesis report with executive summary : a synthesis of the global and sub-global IAASTD reports / edited by Beverly D. McIntyre . . . [et al.].
    p.   cm.
  Includes bibliographical references.
  ISBN 978-1-59726-550-8 (pbk. : alk. paper)
  1. Agriculture—International cooperation. 2. Sustainable development. I. McIntyre, Beverly D. II. Title: Synthesis report with executive summary : a synthesis of the global and sub-global IAASTD reports.
  HD1428.I547 2008
  338.9′27—dc22   2008046049

British Cataloguing-in-Publication data available.

Printed on recycled, acid-free paper

Interior and cover designs by Linda McKnight, McKnight Design, LLC.

Manufactured in the United States of America

10 9 8 7 6 5 4 3 2 1

Contents

Foreword

The objective of the International Assessment of Agricultural Knowledge, Science and Technology for Development (IAASTD) was to assess the impacts of past, present and future agricultural knowledge, science and technology on the:
- reduction of hunger and poverty,
- improvement of rural livelihoods and human health, and
- equitable, socially, environmentally and economically sustainable development.

The IAASTD was initiated in 2002 by the World Bank and the Food and Agriculture Organization of the United Nations (FAO) as a global consultative process to determine whether an international assessment of agricultural knowledge, science and technology was needed. Mr. Klaus Töepfer, Executive Director of the United Nations Environment Programme (UNEP) opened the first Intergovernmental Plenary (30 August-3 September 2004) in Nairobi, Kenya, during which participants initiated a detailed scoping, preparation, drafting and peer review process.

The outputs from this assessment are a Global and five Sub-Global reports; a Global and five Sub-Global Summaries for Decision Makers; and a cross-cutting Synthesis Report with an Executive Summary. The Summaries for Decision Makers and the Synthesis Report specifically provide options for action to governments, international agencies, academia, research organizations and other decision makers around the world.

The reports draw on the work of hundreds of experts from all regions of the world who have participated in the preparation and peer review process. As has been customary in many such global assessments, success depended first and foremost on the dedication, enthusiasm and cooperation of these experts in many different but related disciplines. It is the synergy of these interrelated disciplines that permitted IAASTD to create a unique, interdisciplinary regional and global process.

We take this opportunity to express our deep gratitude to the authors and reviewers of all of the reports—their dedication and tireless efforts made the process a success. We thank the Steering Committee for distilling the outputs of the consultative process into recommendations to the Plenary, the IAASTD Bureau for their advisory role during the assessment and the work of those in the extended Secretariat. We would specifically like to thank the cosponsoring organizations of the Global Environment Facility (GEF) and the World Bank for their financial contributions as well as the FAO, UNEP, and the United Nations Educational, Scientific and Cultural Organization (UNESCO) for their continued support of this process through allocation of staff resources.

We acknowledge with gratitude the governments and organizations that contributed to the Multidonor Trust Fund (Australia, Canada, the European Commission, France, Ireland, Sweden, Switzerland, and the United Kingdom) and the United States Trust Fund. We also thank the governments who provided support to Bureau members, authors and reviewers in other ways. In addition, Finland provided direct support to the Secretariat. The IAASTD was especially successful in engaging a large number of experts from developing countries and countries with economies in transition in its work; the Trust Funds enabled financial assistance for their travel to the IAASTD meetings.

We would also like to make special mention of the Regional Organizations who hosted the regional coordinators and staff and provided assistance in management and time to ensure success of this enterprise: the African Center for Technology Studies (ACTS) in Kenya, the Inter-American Institute for Cooperation on Agriculture (IICA) in Costa Rica, the International Center for Agricultural Research in the Dry Areas (ICARDA) in Syria, and the WorldFish Center in Malaysia.

The final Intergovernmental Plenary in Johannesburg, South Africa was opened on 7 April 2008 by Achim Steiner, Executive Director of UNEP. This Plenary saw the acceptance of the Reports and the approval of the Summaries for Decision Makers and the Executive Summary of the Synthesis Report by an overwhelming majority of governments.

Signed:

Co-chairs
Hans H. Herren
Judi Wakhungu

Director
Robert T. Watson

Preface

In August 2002, the World Bank and the Food and Agriculture Organization (FAO) of the United Nations initiated a global consultative process to determine whether an international assessment of agricultural knowledge, science and technology (AKST) was needed. This was stimulated by discussions at the World Bank with the private sector and nongovernmental organizations (NGOs) on the state of scientific understanding of biotechnology and more specifically transgenics. During 2003, eleven consultations were held, overseen by an international multistakeholder steering committee and involving over 800 participants from all relevant stakeholder groups, e.g., governments, the private sector and civil society. Based on these consultations the steering committee recommended to an Intergovernmental Plenary meeting in Nairobi in September 2004 that an international assessment of the role of AKST in reducing hunger and poverty, improving rural livelihoods and facilitating environmentally, socially and economically sustainable development was needed. The concept of an International Assessment of Agricultural Knowledge, Science and Technology for Development (IAASTD) was endorsed as a multi-thematic, multi-spatial, multi-temporal intergovernmental process with a multistakeholder Bureau cosponsored by the FAO, the Global Environment Facility (GEF), United Nations Development Programme (UNDP), United Nations Environment Programme (UNEP), United Nations Educational, Scientific and Cultural Organization (UNESCO), the World Bank and World Health Organization (WHO).

The IAASTD's governance structure is a unique hybrid of the Intergovernmental Panel on Climate Change (IPCC) and the nongovernmental Millennium Ecosystem Assessment (MA). The stakeholder composition of the Bureau was agreed at the Intergovernmental Plenary meeting in Nairobi; it is geographically balanced and multistakeholder with 30 government and 30 civil society representatives (NGOs, producer and consumer groups, private sector entities and international organizations) in order to ensure ownership of the process and findings by a range of stakeholders.

About 400 of the world's experts were selected by the Bureau, following nominations by stakeholder groups, to prepare the IAASTD Report (comprised of a Global and five Sub-Global assessments). These experts worked in their own capacity and did not represent any particular stakeholder group. Additional individuals, organizations and governments were involved in the peer review process.

The IAASTD development and sustainability goals were endorsed at the first Intergovernmental Plenary and are consistent with a subset of the UN Millennium Development Goals (MDGs): the reduction of hunger and poverty, the improvement of rural livelihoods and human health, and facilitating equitable, socially, environmentally and economically sustainable development. Realizing these goals requires acknowledging the multifunctionality of agriculture: the challenge is to simultaneously meet development and sustainability goals while increasing agricultural production.

Meeting these goals has to be placed in the context of a rapidly changing world of urbanization, growing inequities, human migration, globalization, changing dietary preferences, climate change, environmental degradation, a trend toward biofuels and an increasing population. These conditions are affecting local and global food security and putting pressure on productive capacity and ecosystems. Hence there are unprecedented challenges ahead in providing food within a global trading system where there are other competing uses for agricultural and other natural resources. AKST alone cannot solve these problems, which are caused by complex political and social dynamics, but it can make a major contribution to meeting development and sustainability goals. Never before has it been more important for the world to generate and use AKST.

Given the focus on hunger, poverty and livelihoods, the IAASTD pays special attention to the current situation, issues and potential opportunities to redirect the current AKST system to improve the situation for poor rural people, especially small-scale farmers, rural laborers and others with limited resources. It addresses issues critical to formulating policy and provides information for decision makers confronting conflicting views on contentious issues such as the environmental consequences of productivity increases, environmental and human health impacts of transgenic crops, the consequences of bioenergy development on the environment and on the long-term availability and price of food, and the implications of climate change on agricultural production. The Bureau agreed that the scope of the assessment needed to go beyond the narrow confines of science and technology (S&T) and should encompass other types of relevant knowledge (e.g., knowledge held by agricultural producers, consumers and end users) and that it should also assess the role of institutions, organizations, governance, markets and trade.

The IAASTD is a multidisciplinary and multistakeholder enterprise requiring the use and integration of information, tools and models from different knowledge paradigms including local and traditional knowledge. The IAASTD does not advocate specific policies or practices; it assesses the major issues facing AKST and points towards a range of AKST options for action that meet development and sustainability goals. It is policy relevant, but not policy prescriptive. It integrates scientific information on a range of topics that are critically interlinked, but often addressed independently, i.e., agriculture, poverty, hunger, human health, natural resources, environment, development and innovation. It will enable decision makers to bring a richer base of knowledge to bear on policy and management decisions on issues previously viewed in isolation. Knowledge gained from historical analysis (typically the past 50 years) and an analysis of some future development alternatives to 2050 form the basis for assessing options for action on science and technology, capacity development, institutions and policies, and investments.

The IAASTD is conducted according to an open, transparent, representative and legitimate process; is evidence-based; presents options rather than recommendations; assesses different local, regional and global perspectives; presents different views, acknowledging that there can be more than one interpretation of the same evidence based on different worldviews; and identifies the key scientific uncertainties and areas on which research could be focused to advance development and sustainability goals.

The IAASTD is composed of a Global assessment and five Sub-Global assessments: Central and West Asia and North Africa (CWANA); East and South Asia and the Pacific (ESAP); Latin America and the Caribbean (LAC); North America and Europe (NAE); Sub-Saharan Africa (SSA). It (1) assesses the generation, access, dissemination and use of public and private sector AKST in relation to the goals, using local, traditional and formal knowledge; (2) analyzes existing and emerging technologies, practices, policies and institutions and their impact on the goals; (3) provides information for decision makers in different civil society, private and public organizations on options for improving policies, practices, institutional and organizational arrangements to enable AKST to meet the goals; (4) brings together a range of stakeholders (consumers, governments, international agencies and research organizations, NGOs, private sector, producers, the scientific community) involved in the agricultural sector and rural development to share their experiences, views, understanding and vision for the future; and (5) identifies options for future public and private investments in AKST. In addition, the IAASTD will enhance local and regional capacity to design, implement and utilize similar assessments.

In this assessment agriculture is used to include production of food, feed, fuel, fiber and other products and to include all sectors from production of inputs (e.g., seeds and fertilizer) to consumption of products. However, as in all assessments, some topics were covered less extensively than others (e.g., livestock, forestry, fisheries and the agricultural sector of small island countries, and agricultural engineering), largely due to the expertise of the selected authors. Originally the Bureau approved a chapter on plausible futures (a visioning exercise), but later there was agreement to delete this chapter in favor of a more simple set of model projections. Similarly the Bureau approved a chapter on capacity development, but this chapter was dropped and key messages integrated into other chapters.

The IAASTD draft Report was subjected to two rounds of peer review by governments, organizations and individuals. These drafts were placed on an open access web site and open to comments by anyone. The authors revised the drafts based on numerous peer review comments, with the assistance of review editors who were responsible for ensuring the comments were appropriately taken into account. One of the most difficult issues authors had to address was criticisms that the report was too negative. In a scientific review based on empirical evidence, this is always a difficult comment to handle, as criteria are needed in order to say whether something is negative or positive. Another difficulty was responding to the conflicting views expressed by reviewers. The difference in views was not surprising given the range of stakeholder interests and perspectives. Thus one of the key findings of the IAASTD is that there are diverse and conflicting interpretations of past and current events, which need to be acknowledged and respected.

The Global and Sub-Global Summaries for Decision Makers and the Executive Summary of the Synthesis Report were approved at an Intergovernmental Plenary in April 2008. The Synthesis Report integrates the key findings from the Global and Sub-Global assessments, and focuses on eight Bureau-approved topics: bioenergy; biotechnology; climate change; human health; natural resource management; traditional knowledge and community based innovation; trade and markets; and women in agriculture.

The IAASTD builds on and adds value to a number of recent assessments and reports that have provided valuable information relevant to the agricultural sector, but have not specifically focused on the future role of AKST, the institutional dimensions and the multifunctionality of agriculture. These include: FAO State of Food Insecurity in the World (yearly); InterAcademy Council Report: Realizing the Promise and Potential of African Agriculture (2004); UN Millennium Project Task Force on Hunger (2005); Millennium Ecosystem Assessment (2005); CGIAR Science Council Strategy and Priority Setting Exercise (2006); Comprehensive Assessment of Water Management in Agriculture: Guiding Policy Investments in Water, Food, Livelihoods and Environment (2007); Intergovernmental Panel on Climate Change Reports (2001 and 2007); UNEP Fourth Global Environmental Outlook (2007); World Bank World Development Report: Agriculture for Development (2008); IFPRI Global Hunger Indices (yearly); and World Bank Internal Report of Investments in SSA (2007).

Financial support was provided to the IAASTD by the cosponsoring agencies, the governments of Australia, Canada, Finland, France, Ireland, Sweden, Switzerland, US and UK, and the European Commission. In addition, many organizations have provided in-kind support. The authors and review editors have given freely of their time, largely without compensation.

The Global and Sub-Global Summaries for Decision Makers and the Synthesis Report are written for a range of stakeholders, i.e., government policy makers, private sector, NGOs, producer and consumer groups, international organizations and the scientific community. There are no recommendations, only options for action. The options for action are not prioritized because different options are actionable by different stakeholders, each of whom have a different set of priorities and responsibilities and operate in different socioeconomic and political circumstances.

Executive Summary of the Synthesis Report

Writing team: Tsedeke Abate (Ethiopia), Jean Albergel (France), Inge Armbrecht (Colombia), Patrick Avato (Germany/Italy), Satinder Bajaj (India), Nienke Beintema (the Netherlands), Rym Ben Zid (Tunisia), Rodney Brown (USA), Lorna M. Butler (Canada), Fabrice Dreyfus (France), Kristie L. Ebi (USA), Shelley Feldman (USA), Alia Gana (Tunisia), Tirso Gonzales (Peru), Ameenah Gurib-Fakim (Mauritius), Jack Heinemann (New Zealand), Thora Herrmann (Germany), Angelika Hilbeck (Switzerland), Hans Hurni (Switzerland), Sophia Huyer (Canada), Janice Jiggins (UK), Joan Kagwanja (Kenya), Moses Kairo (Kenya), Rose R. Kingamkono (Tanzania), Gordana Kranjac-Berisavljevic (Ghana), Kawther Latiri (Tunisia), Roger Leakey (Australia), Marianne Lefort (France), Karen Lock (UK), Thora Herrmann (Germany), Yalem Mekonnen (Ethiopia), Douglas Murray (USA), Dev Nathan (India), Lindela Ndlovu (Zimbabwe), Balgis Osman-Elasha (Sudan), Ivette Perfecto (Puerto Rico), Cristina Plencovich (Argentina), Rajeswari Raina (India), Elizabeth Robinson (UK), Niels Roling (Netherlands), Mark Rosegrant (USA), Erika Rosenthal (USA), Wahida Patwa Shah (Kenya), John M.R. Stone (Canada), Abid Suleri (Pakistan), Hong Yang (Australia)

Statement by Governments on Executive Summary

All countries present at the final intergovernmental plenary session held in Johannesburg, South Africa in April 2008 welcome the work of the IAASTD and the uniqueness of this independent multistakeholder and multidisciplinary process, and the scale of the challenge of covering a broad range of complex issues. The Governments present recognize that the Global and Sub-Global Reports are the conclusions of studies by a wide range of scientific authors, experts and development specialists and while presenting an overall consensus on the importance of agricultural knowledge, science and technology for development they also provide a diversity of views on some issues.

All countries see these Reports as a valuable and important contribution to our understanding on agricultural knowledge, science and technology for development recognizing the need to further deepen our understanding of the challenges ahead. This Assessment is a constructive initiative and important contribution that all governments need to take forward to ensure that agricultural knowledge, science and technology fulfils its potential to meet the development and sustainability goals of the reduction of hunger and poverty, the improvement of rural livelihoods and human health, and facilitating equitable, socially, environmentally and economically sustainable development.

In accordance with the above statement, the following governments approve the Executive Summary of the Synthesis Report.

Armenia, Azerbaijan, Bahrain, Bangladesh, Belize, Benin, Bhutan, Botswana, Brazil, Cameroon, People's Republic of China, Costa Rica, Cuba, Democratic Republic of Congo, Dominican Republic, El Salvador, Ethiopia, Finland, France, Gambia, Ghana, Honduras, India, Iran, Ireland, Kenya, Kyrgyzstan, Lao People's Democratic Republic, Lebanon, Libyan Arab Jamahiriya, Maldives, Republic of Moldova, Mozambique, Namibia, Nigeria, Pakistan, Panama, Paraguay, Philippines, Poland, Republic of Palau, Romania, Saudi Arabia, Senegal, Solomon Islands, Swaziland, Sweden, Switzerland, United Republic of Tanzania, Timor-Leste, Togo, Tunisia, Turkey, Uganda, United Kingdom of Great Britain, Uruguay, Viet Nam, Zambia (58 countries).

While approving the above statement the following governments did not fully approve the Executive Summary of the Synthesis Report and their reservations are entered in the Annex to the Executive Summary.

Australia, Canada, United States of America (3 countries).

Executive Summary of the Synthesis Report of the International Assessment of Agricultural Knowledge, Science and Technology for Development (IAASTD)

This Synthesis Report captures the complexity and diversity of agriculture and agricultural knowledge, science and technology (AKST) across world regions. It is built upon the Global and five Sub-Global reports that provide evidence for the integrated analysis of the main concerns necessary to achieve development and sustainability goals. It is organized in two parts that address the primary animating question: how can AKST be used to reduce hunger and poverty, improve rural livelihoods, and facilitate equitable environmentally, socially, and economically sustainable development? In the first part we identify the current conditions, challenges and options for action that shape AKST, while in the second part we focus on eight cross-cutting themes. The eight cross-cutting themes include: bioenergy, biotechnology, climate change, human health, natural resource management, trade and markets, traditional and local knowledge and community-based innovation, and women in agriculture.

The International Assessment of Agricultural Knowledge, Science and Technology for Development (IAASTD) responds to the widespread realization that despite significant scientific and technological achievements in our ability to increase agricultural productivity, we have been less attentive to some of the unintended social and environmental consequences of our achievements. We are now in a good position to reflect on these consequences and to outline various policy options to meet the challenges ahead, perhaps best characterized as the need for food and livelihood security under increasingly constrained environmental conditions from within and outside the realm of agriculture and globalized economic systems.

This widespread realization is linked directly to the goals of the IAASTD: how AKST can be used to reduce hunger and poverty, to improve rural livelihoods and to facilitate equitable environmentally, socially and economically sustainable development. Under the rubric of IAASTD, we recognize the importance of AKST to the multifunctionality of agriculture and the intersection with other local to global concerns, including loss of biodiversity and ecosystem services, climate change and water availability.

The IAASTD is unique in the history of agricultural science assessments in that it assesses both formal science and technology (S&T) and local and traditional knowledge, addresses not only production and productivity but the multifunctionality of agriculture, and recognizes that multiple perspectives exist on the role and nature of AKST. For many years, agricultural science focused on delivering component technologies to increase farm-level productivity where the market and institutional arrangements put in place by the state were the primary drivers of the adoption of new technologies. The general model has been to continuously innovate, reduce farm gate prices and externalize costs. This model drove the phenomenal achievements of AKST in industrial countries after World War II and the spread of the Green Revolution beginning in the 1960s. But, given the new challenges we confront today, there is increasing recognition within formal S&T organizations that the current AKST model requires revision. Business as usual is no longer an option. This leads to rethinking the role of AKST in achieving development and sustainability goals; one that seeks more intensive engagement across diverse worldviews and possibly contradictory approaches in ways that can inform and suggest strategies for actions enabling the multiple functions of agriculture.

In order to address the diverse needs and interests that shape human life, we need a shared approach to sustainability with local and cross-national collaboration. We cannot escape our predicament by simply continuing to rely on the aggregation of individual choices to achieve sustainable and equitable collective outcomes. Incentives are needed to influence the choices individuals make. Issues such as poverty and climate change also require collective agreements on concerted action and governance across scales that go beyond an appeal to individual benefit. At the global, regional, national and local levels, decision makers must be acutely conscious of the fact that there are diverse challenges, multiple theoretical frameworks and development models and a wide range of options to meet development and sustainability goals. Our perception of the challenges and the choices we make at this juncture in history will determine how we protect our planet and secure our future.

Development and sustainability goals should be placed in the context of (1) current social and economic inequities and political uncertainties about war and conflicts; (2) uncertainties about the ability to sustainably produce and access sufficient food; (3) uncertainties about the future of world food prices; (4) changes in the economics of fossil-based energy use; (5) the emergence of new competitors for natural resources; (6) increasing chronic diseases that are partially a consequence of poor nutrition and poor food quality as well as food safety; and (7) changing environmental conditions and the growing awareness of human responsibility for the maintenance of global ecosystem services (provisioning, regulating, cultural and supporting).

Today there is a world of asymmetric development, unsustainable natural resource use, and continued rural and urban poverty. Generally the adverse consequences of global changes have the most significant effects on the poorest and most vulnerable, who historically have had limited entitlements and opportunities for growth.

The pace of formal technology generation and adoption has been highly uneven. Actors within North America and Europe (NAE) and emerging economies who have captured significant economies of scale through formal AKST will continue to dominate agricultural exports and extended value chains. There is an urgent need to diversify and strengthen AKST, recognizing differences in agroecologies and social and cultural conditions. The need to retool AKST, to reduce poverty and provide improved livelihoods options for the rural poor, especially landless and peasant communities, urban, informal and migrant workers, is a major challenge.

There is an overarching concern in all regions regarding poverty alleviation and the livelihoods options available to poor people who are faced with intra- and inter-regional inequalities. There is recognition that the mounting crisis in food security is of a different complexity and potentially different magnitude than the one of the 1960s. The ability and willingness of different actors, including those in the state, civil society and private sector, to address fundamental questions of relationships among production, social and environmental systems is affected by contentious political and economic stances.

The acknowledgment of current challenges and the acceptance of options available for action require a long-term commitment from decision makers that is responsive to the specific needs of a wide range of stakeholders. A recognition that knowledge systems and human ingenuity in science, technology, practice and policy is needed to meet the challenges, opportunities and uncertainties ahead. This recognition will require a shift to nonhierarchical development models.

The main challenge of AKST is to increase the productivity of agriculture in a sustainable manner. AKST must address the needs of small-scale farms in diverse ecosystems and create realistic opportunities for their development where the potential for improved area productivity is low and where climate change may have its most adverse consequences. The main challenges for AKST posed by multifunctional agricultural systems include:

- How to improve social welfare and personal livelihoods in the rural sector and enhance multiplier effects of agriculture?
- How to empower marginalized stakeholders to sustain the diversity of agriculture and food systems, including their cultural dimensions?
- How to provide safe water, maintain biodiversity, sustain the natural resource base and minimize the adverse impacts of agricultural activities on people and the environment?
- How to maintain and enhance environmental and cultural services while increasing sustainable productivity and diversity of food, fiber and biofuel production?
- How to manage effectively the collaborative generation of knowledge among increasingly heterogeneous contributors and the flow of information among diverse public and private AKST organizational arrangements?
- How to link the outputs from marginalized, rain fed lands into local, national and global markets?

Multifunctionality

The term *multifunctionality* has sometimes been interpreted as having implications for trade and protectionism. This is *not* the definition used here. In IAASTD, multifunctionality is used solely to express the inescapable interconnectedness of agriculture's different roles and functions. The concept of multifunctionality recognizes agriculture as a multi-output activity producing not only commodities (food, feed, fibers, agrofuels, medicinal products and ornamentals), but also non-commodity outputs such as environmental services, landscape amenities and cultural heritages.

The working definition proposed by OECD, which is used by the IAASTD, associates multifunctionality with the particular characteristics of the agricultural production process and its outputs; (1) multiple commodity and non-commodity outputs are jointly produced by agriculture; and (2) some of the non-commodity outputs may exhibit the characteristics of externalities or public goods, such that markets for these goods function poorly or are nonexistent.

The use of the term has been controversial and contested in global trade negotiations, and it has centered on whether "trade-distorting" agricultural subsidies are needed for agriculture to perform its many functions. Proponents argue that current patterns of agricultural subsidies, international trade and related policy frameworks do not stimulate transitions toward equitable agricultural and food trade relation or sustainable food and farming systems and have given rise to perverse impacts on natural resources and agroecologies as well as on human health and nutrition. Opponents argue that attempts to remedy these outcomes by means of trade-related instruments will weaken the efficiency of agricultural trade and lead to further undesirable market distortion; their preferred approach is to address the externalized costs and negative impacts on poverty, the environment, human health and nutrition by other means.

Options for Action

Successfully meeting development and sustainability goals and responding to new priorities and changing circumstances would require a fundamental shift in AKST, including science, technology, policies, institutions, capacity development and investment. Such a shift would recognize and give increased importance to the multifunctionality of agriculture, accounting for the complexity of agricultural systems within diverse social and ecological contexts. It would require new institutional and organizational arrangements to promote an integrated approach to the development and deployment of AKST. It would also recognize farming communities, farm households, and farmers as producers <u>and</u> managers of ecosystems. This shift may call for changing the incentive systems for all actors along the value chain to internalize as many externalities as possible. In terms of development and sustainability goals, these policies and institutional changes should be directed primarily at those who have been served least by previous AKST approaches, i.e., resource-poor farmers, women and ethnic minorities.[1] Such development would depend also on the extent to which small-scale farmers can find gainful off-farm employment and help fuel general economic growth. Large and middle-size farmers continue to be important and high pay-off targets of AKST, especially in the area of sustainable land use and food systems.

It will be important to assess the potential environmental, health and social impacts of any technology, and to implement the appropriate regulatory frameworks. AKST can contribute to radically improving food security and enhancing the social and economic performance of agricultural systems as a basis for sustainable rural and community livelihoods and wider economic development. It can help to rehabilitate degraded land, reduce environmental and health risks associated with food production and consumption and sustainably increase production.

Success would require increased public and private investment in AKST, the development of supporting policies and institutions, revalorization of traditional and local knowledge, and an interdisciplinary, holistic and systems-based approach to knowledge production and sharing. Success also depends on the extent to which international developments and events drive the priority given to development and sustainability goals and the extent to which requisite funding and qualified staff are available.

Poverty and livelihoods

Important options for enhancing rural livelihoods include increasing access by small-scale farmers to land and economic resources and to remunerative local urban and export markets; and increasing local value added and value captured by small-scale farmers and rural laborers. A powerful tool for meeting development and sustainability goals resides in empowering farmers to innovatively manage soils, water, biological resources, pests, disease vectors, genetic diversity, and conserve natural resources in a culturally appropriate manner. Combining farmers' and external knowledge would require new partnerships among farmers, scientists and other stakeholders.

Policy options for improving livelihoods include access to microcredit and other financial services; legal frameworks that ensure access and tenure to resources and land; recourse to fair conflict resolution; and progressive evolution and proactive engagement in intellectual property rights (IPR) regimes and related instruments.[2] Developments are needed that build trust and that value farmer knowledge, agricultural and natural biodiversity; farmer-managed medicinal plants, local seed systems and common pool resource management regimes. Each of these options, when implemented locally, depends on regional and nationally based mechanisms to ensure accountability. The suite of options to increase domestic farm gate prices for small-scale farmers includes fiscal and competition policies; improved access to AKST; novel business approaches; and enhanced political power.

---

[1] Botswana.
[2] USA.

> *Food security* [is] a situation that exists when all people, at all times, have physical, *social* and economic access to sufficient, safe and nutritious food that meets their dietary needs and food preferences for an active and healthy life. (FAO, The State of Food Insecurity, 2001)
>
> *Food sovereignty* is defined as the right of peoples and sovereign states to democratically determine their own agricultural and food policies.[3]
>
> ---
>
> [3] UK.

Food security

Food security strategies require a combination of AKST approaches, including the development of food stock management, effective market intelligence and early warning, monitoring, and distribution systems. Production measures create the conditions for food security, but they need to be looked at in conjunction with people's access to food (through own production, exchange and public entitlements) and their ability to absorb nutrients consumed (through adequate access to water and sanitation, adequate nutrition and nutritional information) in order to fully achieve food security.

AKST can increase sustainable agricultural production by expanding use of local and formal AKST to develop and deploy suitable cultivars adaptable to site-specific conditions; improving access to resources; improving soil, water and nutrient management and conservation; pre- and post-harvest pest management; and increasing small-scale farm diversification. Policy options for addressing food security include developing high-value and underutilized crops in rain fed areas; increasing the full range of agricultural exports and imports, including organic and fair trade products; reducing transaction costs for small-scale producers; strengthening local markets; food safety nets; promoting agro-insurance; and improving food safety and quality. Price shocks and extreme weather events call for a global system of monitoring and intervention for the timely prediction of major food shortages and price-induced hunger.

AKST investments can increase the sustainable productivity of major subsistence foods including orphan and underutilized crops, which are often grown or consumed by poor people. Investments could also be targeted for institutional change and policies that can improve access of poor people to food, land, water, seeds, germplasm and improved technologies.

Environmental sustainability

AKST systems are needed that enhance sustainability while maintaining productivity in ways that protect the natural resource base and ecological provisioning of agricultural systems. Options include improving nutrient, energy, water and land use efficiency; improving the understanding of soil-plant-water dynamics; increasing farm diversification;

supporting agroecological systems, and enhancing biodiversity conservation and use at both field and landscape scales; promoting the sustainable management of livestock, forest and fisheries; improving understanding of the agroecological functioning of mosaics of crop production areas and natural habitats; countering the effects of agriculture on climate change and mitigating the negative impacts of climate change on agriculture.

Policy options include ending subsidies that encourage unsustainable practices and using market and other mechanisms to regulate and generate rewards for agro/environmental services, for better natural resource management and enhanced environmental quality. Examples include incentives to promote integrated pest management (IPM) and environmentally resilient germplasm management, payments to farmers and local communities for ecosystem services, facilitating and providing incentives for alternative markets such as green products, certification for sustainable forest and fisheries practices and organic agriculture and the strengthening of local markets. Long-term land and water use rights/tenure, risk reduction measures (safety nets, credit, insurance, etc.) and profitability of recommended technologies are prerequisites for adoption of sustainable practices. Common pool resource regimes and modes of governance that emphasize participatory and democratic approaches are needed.

Investment opportunities in AKST that could improve sustainability and reduce negative environmental effects include resource conservation technologies, improved techniques for organic and low-input systems; a wide range of breeding techniques for temperature and pest tolerance; research on the relationship of agricultural ecosystem services and human well-being; economic and non-economic valuations of ecosystem services; increasing water use efficiency and reducing water pollution; biocontrols of current and emerging pests and pathogens; biological substitutes for agrochemicals; and reducing the dependency of the agricultural sector on fossil fuels.

Human health and nutrition

Inter-linkages between health, nutrition, agriculture, and AKST affect the ability of individuals, communities, and nations to reach sustainability goals. These inter-linkages exist within the context of multiple stressors that affect population health. A broad and integrated approach is needed to identify appropriate use of AKST to increase food security and safety, decrease the incidence and prevalence of a range of infectious (including emerging and reemerging diseases such as malaria, avian influenza, HIV/AIDS and others) and chronic diseases, and decrease occupational exposures, injuries and deaths. Robust agricultural, public health, and veterinary detection, surveillance, monitoring, and response systems can help identify the true burden of ill health and cost-effective, health-promoting strategies and measures. Additional investments are needed to maintain and improve current systems and regulations.

- *Increasing food security* can be facilitated by promoting policies and programs to diversify diets and improve micronutrient intake; and developing and deploying existing and new technologies for the production, processing, preservation, and distribution of food.

- *Increasing food safety* can be facilitated by effective, coordinated, and proactive national and international food safety systems to ensure animal, plant, and human health, such as investments in adequate infrastructure, public health and veterinary capacity, legislative frameworks for identification and control of biological and chemical hazards, and farmer-scientist partnerships for the identification, monitoring and evaluation of risks.

- *The burden of infectious disease* can be decreased by strengthening coordination between and the capacity of agricultural, veterinary, and public health systems; integrating multi-sectoral policies and programs across the food chain to reduce the spread of infectious diseases; and developing and deploying new AKST to identify, monitor, control, and treat diseases.

- *The burden of chronic disease* can be decreased by policies that explicitly recognize the importance of improving human health and nutrition, including regulation of food product formulation through legislation, international agreements and regulations for food labeling and health claims, and creation of incentives for the production and consumption of health-promoting foods.

- *Occupational and public health* can be improved by development and enforcement of health and safety regulations (including child labor laws and pesticide regulations), enforcement of cross-border issues such as illegal use of toxic agrochemicals, and conducting health risk assessments that make explicit the tradeoffs between maximizing livelihood benefits, the environment, and improving health.

Equity

For AKST to contribute to greater equity, investments are required for the development of context-specific technologies, and expanded access of farmers and other rural people to occupational, non-formal and formal education. An environment in which formal science and technology and local and traditional knowledge are seen as part of an integral AKST system can increase equitable access to technologies for a broad range of producers and natural resource managers. Incentives in science, universities and research organizations are needed to foster different kinds of AKST partnerships. Key options include equitable access to and use of natural resources (particularly land and water), systems of incentives and rewards for multifunctionality, including ecosystem services, and responding to the vulnerability of farming and farm worker communities. Reform of the governance of AKST and related organizations is also important for the crucial role they can play in improving community-level scientific literacy, decentralization of technological opportunities, and the integration of farmer concerns in research priority setting and the design of farmer services. Improving equity requires synergy among various development actors, including farmers, rural laborers, banks, civil society organizations, commercial companies, and public agencies. Stakeholder involvement is also crucial in decisions about IPR, infrastructure, tariffs, and the internalization of social and environmental costs. New modes of governance to develop innovative local networks and decentralized government, focusing on small-scale producers and the urban poor (urban agriculture; direct links between urban consumers and rural producers) will help create and strengthen synergistic and complementary capacities.

Preferential investments in equitable development (e.g., literacy, education and training) that contribute to reducing ethnic, gender, and other inequities would advance development goals. Measurements of returns to investments require indices that give more information than GDP, and that are sensitive to environmental and equity gains. The use of inequality indices for screening AKST investments and monitoring outcomes strengthens accountability. The Gini-coefficient could, for example, become a public criterion for policy assessment, in addition to the more conventional measures of growth, inflation and environment.

Investments
Achieving development and sustainability goals would entail increased funds and more diverse funding mechanisms for agricultural research and development and associated knowledge systems, such as:

- Public investments in global, regional, national and local public goods; food security and safety, climate change and sustainability. More efficient use of increasingly scarce land, water and biological resources requires investment in research and development of legal and management capabilities.
- Public investments in agricultural knowledge systems to promote interactive knowledge networks (farmers, scientists, industry and actors in other knowledge areas); improved access to information and communication technologies (ICT); ecological, evolutionary, food, nutrition, social and complex systems' sciences; effective interdisciplinarity; capacity in core agricultural sciences; and improving life-long learning opportunities along the food system.
- Public-private partnerships for improved commercialization of applied knowledge and technologies and joint funding of AKST, where market risks are high and where options for widespread utilization of knowledge exist.
- Adequate incentives and rewards to encourage private and civil society investments in AKST contributing to development and sustainability goals.
- In many developing countries, it may be necessary to complement these investments with increased and more targeted investments in rural infrastructure, education and health.

In the face of new global challenges, there is an urgent need to strengthen, restructure and possibly establish new intergovernmental, independent science and evidence-based networks to address such issues as climate forecasting for agricultural production; human health risks from emerging diseases; reorganization of livelihoods in response to changes in agricultural systems (population movements); food security; and global forestry resources.

Themes
The Synthesis Report looked at eight AKST-related themes of critical interest to meeting development and sustainability goals: bioenergy, biotechnology, climate change, human health, natural resource management, trade and markets, traditional and local knowledge and community-based innovation and women in agriculture.

Bioenergy
Rising costs of fossil fuels, energy security concerns, increased awareness of climate change and potentially positive effects for economic development have led to considerable public attention to bioenergy. Bioenergy includes traditional bioenergy, biomass to produce electricity, light and heat and first and next generation liquid biofuels. The economics and the positive and negative social and environmental externalities differ widely, depending on source of biomass, type of conversion technology and local circumstances.

Primarily due to a lack of affordable alternatives, millions of people in developing countries depend on traditional bioenergy (e.g., wood fuels) for their cooking and heating needs, especially in sub-Saharan Africa and South Asia. This reliance on traditional bioenergy can pose considerable environmental, health, economic and social challenges. New efforts are needed to improve traditional bioenergy and accelerate the transition to more sustainable forms of energy.

First generation biofuels consist predominantly of bioethanol and biodiesel produced from agricultural crops (e.g., maize, sugar cane). Production has been growing fast in recent years, primarily due to biofuel support policies since they are cost competitive only under particularly favorable circumstances. The diversion of agricultural crops to fuel can raise food prices and reduce our ability to alleviate hunger throughout the world. The negative social effects risk being exacerbated in cases where small-scale farmers are marginalized or displaced from their land. From an environmental perspective, there is considerable variation, uncertainty and debate over the net energy balance and level of greenhouse gas (GHG) emissions. In the long term, effects on food prices may be reduced, but environmental effects caused by land and water requirements of large-scale increases of first generation biofuels production are likely to persist and will need to be addressed.

Next generation biofuels such as cellulosic ethanol and biomass-to-liquids technologies allow conversion into biofuels of more abundant and cheaper feedstocks than first generation. This could potentially reduce agricultural land requirements per unit of energy produced and improve life-cycle GHG emissions, potentially mitigating the environmental pressures from first generation biofuels. However, next generation biofuels technologies are not yet commercially proven and environmental and social effects are still uncertain. For example, the use of feedstock and farm residues can compete with the need to maintain organic matter in sustainable agroecosystems.

Bioelectricity and bioheat are important forms of renewable energy that are usually more efficient and produce less GHG emissions than liquid biofuels and fossil fuels. Digesters, gasifiers and direct combustion devices can be successfully employed in certain settings, e.g., off-grid areas. There is potential for expanding these applications but AKST is needed to reduce costs and improve operational reliability. For all forms of bioenergy, decision makers should carefully weigh full social, environmental and economic costs against realistically achievable benefits and other sustainable energy options.

Biotechnology [4]

The IAASTD definition of biotechnology is based on that in the Convention on Biological Diversity and the Cartagena Protocol on Biosafety. It is a broad term embracing the manipulation of living organisms and spans the large range of activities from conventional techniques for fermentation and plant and animal breeding to recent innovations in tissue culture, irradiation, genomics and marker-assisted breeding (MAB) or marker assisted selection (MAS) to augment natural breeding. Some of the latest biotechnologies ("modern biotechnology") include the use of *in vitro* modified DNA or RNA and the fusion of cells from different taxonomic families, techniques that overcome natural physiological reproductive or recombination barriers. Currently the most contentious issue is the use of recombinant DNA techniques to produce transgenes that are inserted into genomes. Even newer techniques of modern biotechnology manipulate heritable material without changing DNA.

Biotechnology has always been on the cutting edge of change. Change is rapid, the domains involved are numerous, and there is a significant lack of transparent communication among actors. Hence assessment of modern biotechnology is lagging behind development; information can be anecdotal and contradictory, and uncertainty on benefits and harms is unavoidable. There is a wide range of perspectives on the environmental, human health and economic risks and benefits of modern biotechnology; many of these risks are as yet unknown.

Conventional biotechnologies, such as breeding techniques, tissue culture, cultivation practices and fermentation are readily accepted and used. Between 1950 and 1980, prior to the development of genetically modified organisms (GMOs), modern varieties of wheat increased yields up to 33% even in the absence of fertilizer. Modern biotechnologies used in containment have been widely adopted; e.g., the industrial enzyme market reached US$1.5 billion in 2000. The application of modern biotechnology outside containment, such as the use of genetically modified (GM) crops is much more contentious. For example, data based on some years and some GM crops indicate highly variable 10-33% yield gains in some places and yield declines in others.

Higher level drivers of biotechnology R&D, such as IPR frameworks, determine what products become available. While this attracts investment in agriculture, it can also concentrate ownership of agricultural resources. An emphasis on modern biotechnology without ensuring adequate support for other agricultural research can alter education and training programs and reduce the number of professionals in other core agricultural sciences. This situation can be self-reinforcing since today's students define tomorrow's educational and training opportunities.

The use of patents for transgenes introduces additional issues. In developing countries especially, instruments such as patents may drive up costs, restrict experimentation by the individual farmer or public researcher while also potentially undermining local practices that enhance food security and economic sustainability. In this regard, there is particular concern about present IPR instruments eventually inhibiting seed-saving, exchange, sale and access to proprietary materials necessary for the independent research community to conduct analyses and long term experimentation on impacts. Farmers face new liabilities: GM farmers may become liable for adventitious presence if it causes loss of market certification and income to neighboring organic farmers, and conventional farmers may become liable to GM seed producers if transgenes are detected in their crops.

A problem-oriented approach to biotechnology research and development (R&D) would focus investment on local priorities identified through participatory and transparent processes, and favor multifunctional solutions to local problems. These processes require new kinds of support for the public to critically engage in assessments of the technical, social, political, cultural, gender, legal, environmental and economic impacts of modern biotechnology. Biotechnologies should be used to maintain local expertise and germplasm so that the capacity for further research resides within the local community. Such R&D would put much needed emphasis onto participatory breeding projects and agroecology.

Climate change

Climate change, which is taking place at a time of increasing demand for food, feed, fiber and fuel, has the potential to irreversibly damage the natural resource base on which agriculture depends. The relationship between climate change and agriculture is a two-way street; agriculture contributes to climate change in several major ways and climate change in general adversely affects agriculture.

In mid- to high-latitude regions moderate local increases in temperature can have small beneficial impacts on crop yields; in low-latitude regions, such moderate temperature increases are likely to have negative yield effects. Some negative impacts are already visible in many parts of the world; additional warming will have increasingly negative impacts in all regions. Water scarcity and the timing of water availability will increasingly constrain production. Climate change will require a new look at water storage to cope with the impacts of more and extreme precipitation, higher intra- and inter-seasonal variations, and increased rates of evapotranspiration in all types of ecosystems. Extreme climate events (floods and droughts) are increasing and expected to amplify in frequency and severity and there are likely to be significant consequences in all regions for food and forestry production and food insecurity. There is a serious potential for future conflicts over habitable land and natural resources such as freshwater. Climate change is affecting the distribution of plants, invasive species, pests and disease vectors and the geographic range and incidence of many human, animal and plant diseases is likely to increase.

A comprehensive approach with an equitable regulatory framework, differentiated responsibilities and intermediate targets are required to reduce GHG emissions. The earlier and stronger the cuts in emissions, the quicker concentrations will approach stabilization. Emission reduction measures clearly are essential because they can have an impact

---

[4] China and USA.

due to inertia in the climate system. However, since further changes in the climate are inevitable adaptation is also imperative. Actions directed at addressing climate change and promoting sustainable development share some important goals such as equitable access to resources and appropriate technologies.

Some "win-win" mitigation opportunities have already been identified. These include land use approaches such as lower rates of agricultural expansion into natural habitats; afforestation, reforestation, increased efforts to avoid deforestation, agroforestry, agroecological systems, and restoration of underutilized or degraded lands and rangelands and land use options such as carbon sequestration in agricultural soils, reduction and more efficient use of nitrogenous inputs; effective manure management and use of feed that increases livestock digestive efficiency. Policy options related to regulations and investment opportunities include financial incentives to maintain and increase forest area through reduced deforestation and degradation and improved management and the development and utilization of renewable energy sources. The post-2012 regime has to be more inclusive of all agricultural activities such as reduced emission from deforestation and soil degradation to take full advantage of the opportunities offered by agriculture and forestry sectors.

Human health

Despite the evident and complex links between health, nutrition, agriculture, and AKST, improving human health is not generally an explicit goal of agricultural policy. Agriculture and AKST can affect a range of health issues including undernutrition, chronic diseases, infectious diseases, food safety, and environmental and occupational health. Ill heath in the farming community can in turn reduce agricultural productivity and the ability to develop and deploy appropriate AKST. Ill health can result from undernutrition, as well as over-nutrition. Despite increased global food production over recent decades, undernutrition is still a major global public health problem, causing over 15% of the global disease burden. Protein energy and micronutrient malnutrition remain challenges, with high variability between and within countries. Food security can be improved through policies and programs to increase dietary diversity and through development and deployment of existing and new technologies for production, processing, preservation, and distribution of food.

AKST policies and practices have increased production and new mechanisms for food processing. Reduced dietary quality and diversity and inexpensive foods with low nutrient density have been associated with increasing rates of worldwide obesity and chronic disease. Poor diet throughout the life course is a major risk factor for chronic diseases, which are the leading cause of global deaths. There is a need to focus on consumers and the importance of dietary quality as main drivers of production, and not merely on quantity or price. Strategies include fiscal policies (taxation, trade regimes) for health-promoting foods and regulation of food product formulation, labeling and commercial information.

Globalization of the food supply, accompanied by concentration of food distribution and processing companies, and growing consumer awareness increase the need for effective, coordinated, and proactive national food safety systems. Health concerns that could be addressed by AKST include the presence of pesticide residues, heavy metals, hormones, antibiotics and various additives in the food system as well as those related to large-scale livestock farming.

Strengthened food safety measures are important and necessary in both domestic and export markets and can impose significant costs. Some countries may need help in meeting food control costs such as monitoring and inspection, and costs associated with market rejection of contaminated commodities. Taking a broad and integrated agroecosystem and human health approach can facilitate identification of animal, plant, and human health risks, and appropriate AKST responses.

Worldwide, agriculture accounts for at least 170,000 occupational deaths each year: half of all fatal accidents. Machinery and equipment, such as tractors and harvesters, account for the highest rates of injury and death, particularly among rural laborers. Other important health hazards include agrochemical poisoning, transmissible animal diseases, toxic or allergenic agents, and noise, vibration and ergonomic hazards. Improving occupational health requires a greater emphasis on health protection through development and enforcement of health and safety regulations. Policies should explicitly address tradeoffs between livelihood benefits and environmental, occupational and public health risks.

The incidence and geographic range of many emerging and reemerging infectious diseases are influenced by the intensification of crop and livestock systems. Serious socioeconomic impacts can arise when diseases spread widely within human or animal populations, or when they spill over from animal reservoirs to human hosts. Most of the factors that contribute to disease emergence will continue, if not intensify. Integrating policies and programs across the food chain can help reduce the spread of infectious diseases; robust detection, surveillance, monitoring, and response programs are critical.

Natural resource management [5]

Natural resources, especially those of soil, water, plant and animal diversity, vegetation cover, renewable energy sources, climate and ecosystem services are fundamental for the structure and function of agricultural systems and for social and environmental sustainability, in support of life on earth. Historically the path of global agricultural development has been narrowly focused on increased productivity rather than on a more holistic integration of natural resources management (NRM) with food and nutritional security. A holistic, or systems-oriented approach, is preferable because it can address the difficult issues associated with the complexity of food and other production systems in different ecologies, locations and cultures.

AKST to resolve NRM exploitation issues, such as the mitigation of soil fertility through synthetic inputs and natural processes, is often available and well understood.

---

[5] Capture fisheries and forestry have not been as well covered as other aspects of NRM.

Nevertheless, the resolution of natural resource challenges will demand new and creative approaches by stakeholders with diverse backgrounds, skills and priorities. Capabilities for working together at multiple scales and across different social and physical environments are not well developed. For example, there have been few opportunities for two-way learning between farmers and researchers or policy makers. Consequently farmers and civil society members have seldom been involved in shaping NRM policy. Community-based partnerships with the private sector, now in their early stages of development, represent a new and promising way forward.

The following high priority NRM options for action are proposed:

- Use existing AKST to identify and address some of the underlying causes of declining productivity embedded in natural resource mismanagement, and develop new AKST based on multidisciplinary approaches for a better understanding of the complexity in NRM. Part of this process will involve the cost-effective monitoring of trends in the utilization of natural resource capital.
- Strengthen human resources in the support of natural capital through increased investment (research, training and education, partnerships, policy) in promoting the awareness of the societal costs of degradation and value of ecosystems services.
- Promote research "centers of AKST-NRM excellence" to facilitate less exploitative NRM and better strategies for resource resilience, protection and renewal through innovative two-way learning processes in research and development, monitoring and policy formulation.
- Create an enabling environment for building NRM capacity and increasing understanding of NRM among stakeholders and their organizations in order to shape NRM policy in partnership with public and private sectors.
- Develop networks of AKST practitioners (farmer organizations, NGOs, government, private sector) to facilitate long-term natural resource management to enhance benefits from natural resources for the collective good.
- Connect globalization and localization pathways that link locally generated NRM knowledge and innovations to public and private AKST.

When AKST is developed and used creatively with active participation among various stakeholders across multiple scales, the misuse of natural capital can be reversed and the judicious use and renewal of water bodies, soils, biodiversity, ecosystems services, fossil fuels and atmospheric quality ensured for future generations.

Trade and markets

Targeting market and trade policies to enhance the ability of agricultural and AKST systems to drive development, strengthen food security, maximize environmental sustainability, and help make the small-scale farm sector profitable to spearhead poverty reduction is an immediate challenge around the world.

Agricultural trade can offer opportunities for the poor, but current arrangements have major distributional impacts among, and within, countries that in many cases have not been favorable for small-scale farmers and rural livelihoods. These distributional impacts call for differentiation in policy frameworks and institutional arrangements if these countries are to benefit from agricultural trade. There is growing concern that opening national agricultural markets to international competition before basic institutions and infrastructure are in place can undermine the agricultural sector, with long-term negative effects for poverty, food security and the environment.[6]

Trade policy reform to provide a fairer global trading system can make a positive contribution to sustainability and development goals. Special and differential treatment accorded through trade negotiations can enhance the ability of developing countries to pursue food security and development goals while minimizing trade-related dislocations. Preserving national policy flexibility allows developing countries to balance the needs of poor consumers (urban and rural landless) and rural small-scale farmers. Increasing the value captured by small-scale farmers in global, regional and local markets chains is fundamental to meeting development and sustainability goals. Supportive trade policies can also make new AKST available to the small-scale farm sector and agroenterprises.

Developing countries would benefit from the removal of barriers for products in which they have a comparative advantage; reduction of escalating tariffs for processed commodities in industrialized and developing countries; deeper preferential access to markets for least developed countries; increased public investment in rural infrastructure and the generation of public goods AKST; and improved access to credit, AKST resources and markets for poor producers. Compensating revenues lost as a result of tariff reductions is essential to advancing development agendas.[7]

Agriculture generates large environmental externalities, many of which derive from failure of markets to value environmental and social harm and provide incentives for sustainability. AKST has great potential to reverse this trend. Market and trade policies to facilitate the contribution of AKST to reducing the environmental footprint of agriculture include removing resource use–distorting subsidies; taxing externalities; better definitions of property rights; and developing rewards and markets for agroenvironmental services, including the extension of carbon financing, to provide incentives for sustainable agriculture.

The quality and transparency of governance in the agricultural sector, including increased participation of stakeholders in AKST decision making is fundamental. Strengthening developing country trade analysis and negotiation capacity, and providing better tools for assessing tradeoffs in proposed trade agreements are important to improving governance.

Traditional and local knowledge and community-based innovation

Once AKST is directed simultaneously toward production, profitability, ecosystem services and food systems that are site-specific and evolving, then formal, traditional and lo-

---

[6] USA.
[7] Canada and USA.

cal knowledge need to be integrated. Traditional and local knowledge constitutes an extensive realm of accumulated practical knowledge and knowledge-generating capacity that is needed if sustainability and development goals are to be reached. The traditional knowledge, identities and practices of indigenous and local communities are recognized under the UN Convention on Biological Diversity as embodying ways of life relevant for conservation and sustainable use of biodiversity; and by others as generated by the purposeful interaction of material and non-material worlds embedded in place-based cultures and identities. Local knowledge refers to capacities and activities that exist among rural people in all parts of the world.

Traditional and local knowledge is dynamic; it may sometimes fail but also has had well-documented, extensive, positive impacts. Participatory collaboration in knowledge generation, technology development and innovation has been shown to add value to science-based technology development, for instance in Farmer-Researcher groups in the Andes, in Participatory Plant Breeding, the domestication of wild and semiwild tree species and in soil and water management.

Options for action with proven contribution to achieving sustainability and development goals include collaboration in the conservation, development and use of local and traditional biological materials; incentives for and development of capacity among scientists and formal research organizations to work with local and indigenous people and their organizations; a higher profile in scientific education for indigenous and local knowledge as well as for professional and community-based archiving and assessment of such knowledge and practices. The role of modern ICT in achieving effective collaboration is critical to evolving culturally appropriate integration and merits larger investments and support. Effective collaboration and integration would be supported by international intellectual property and other regimes that allow more scope for dealing effectively with situations involving traditional knowledge, genetic resources and community-based innovations. Examples of misappropriation of indigenous and local people's knowledge and community-based innovations indicate a need for sharing of information about existing national *sui generis* and regulatory frameworks.

Women in agriculture
Gender, that is socially constructed relations between men and women, is an organizing element of existing farming systems worldwide and a determining factor of ongoing agricultural restructuring. Current trends in agricultural market liberalization and in the reorganization of farm work, as well as the rise of environmental and sustainability concerns are redefining the links between gender and development. The proportion of women in agricultural production and postharvest activities ranges from 20 to 70%; their involve-

ment is increasing in many developing countries, particularly with the development of export-oriented irrigated farming, which is associated with a growing demand for female labor, including migrant workers.

Whereas these dynamics have in some ways brought benefits, in general, the largest proportion of rural women worldwide continues to face deteriorating health and work conditions, limited access to education and control over natural resources, insecure employment and low income. This situation is due to a variety of factors, including the growing competition on agricultural markets which increases the demand for flexible and cheap labor, growing pressure on and conflicts over natural resources, the diminishing support by governments for small-scale farms and the reallocation of economic resources in favor of large agroenterprises. Other factors include increasing exposure to risks related to natural disasters and environmental changes, worsening access to water, increasing occupational and health risks.

Despite progress made in national and international policies since the first world conference on women in 1975, urgent action is still necessary to implement gender and social equity in AKST policies and practices if we are to better address gender issues as integral to development processes. Such action includes strengthening the capacity of public institutions and NGOs to improve the knowledge of women's changing forms of involvement in farm and other rural activities in AKST. It also requires giving priority to women's access to education, information, science and technology, and extension services to enable improving women's access, ownership and control of economic and natural resources. To ensure such access, ownership and control legal measures, appropriate credit schemes, support for women's income generating activities and the reinforcement of women's organizations and networks are needed. This, in turn, depends on strengthening women's ability to benefit from market-based opportunities by institutions and policies giving explicit priority to women farmer groups in value chains.

A number of other changes will strengthen women's contributions to agricultural production and sustainability. These include support for public services and investment in rural areas in order to improve women's living and working conditions; giving priority to technological development policies targeting rural and farm women's needs and recognizing their knowledge, skills and experience in the production of food and the conservation of biodiversity; and assessing the negative effects and risks of farming practices and technology, including pesticides on women's health, and taking measures to reduce use and exposure. Finally, if we are to better recognize women as integral to sustainable development, it is critical to ensure gender balance in AKST decision-making at all levels and provide mechanisms to hold AKST organizations accountable for progress in the above areas.

Reservations on Executive Summary

Reservations on full Executive Summary

Australia: Australia recognizes the IAASTD initiative and reports as a timely and important multistakeholder and multidisciplinary exercise designed to assess and enhance the role of AKST in meeting the global development challenges. The wide range of observations and views presented however, are such that Australia cannot agree with all assertions and options in the report. The report is therefore noted as a useful contribution which will be used for considering the future priorities and scope of AKST in securing economic growth and the alleviation of hunger and poverty.

Canada: The Canadian Government recognizes the significant work undertaken by IAASTD authors, Secretariat and stakeholders and notes the Executive Summary of the Synthesis Report as a valuable and important contribution to policy debate which needs to continue in national and international processes. While acknowledging considerable improvement has been achieved through a process of compromise, there remain a number of assertions and observations that require more substantial, balanced and objective analysis. However, the Canadian Government advocates it be drawn to the attention of governments for consideration in addressing the importance of AKST and its large potential to contribute to economic growth and the reduction of hunger and poverty.

United States of America: The United States joins consensus with other governments in the critical importance of AKST to meet the goals of the IAASTD. We commend the tireless efforts of the authors, editors, Co-Chairs and the Secretariat. We welcome the IAASTD for bringing together the widest array of stakeholders for the first time in an initiative of this magnitude. We respect the wide diversity of views and healthy debate that took place.

As we have specific and substantive concerns in each of the reports, the United States is unable to provide unqualified endorsement of the reports, and we have noted them.

The United States believes the Assessment has potential for stimulating further deliberation and research. Further, we acknowledge the reports are a useful contribution for consideration by governments of the role of AKST in raising sustainable economic growth and alleviating hunger and poverty.

Reservations on individual passages

1. Botswana notes that this is specially a problem in sub-Saharan Africa.
2. The USA would prefer that this sentence be written as follows "progressive evolution of IPR regimes in countries where national policies are not fully developed and progressive engagement in IPR management."
3. The UK notes that there is no international definition of food sovereignty.
4. China and USA do not believe that this entire section is balanced and comprehensive.
6. The USA would prefer that this sentence be reflected in this paragraph: "Opening national agricultural markets to international competition can offer economic benefits, but can lead to long-term negative effects on poverty alleviation, food security and the environment without basic national institutions and infrastructure being in place."
7. Canada and USA would prefer the following sentence: "Provision of assistance to help low income countries affected by liberalization to adjust and benefit from liberalized trade is essential to advancing development agendas."

Synthesis Report
A Synthesis of the Global and Sub-Global IAASTD Reports

Statement by Governments on Synthesis Report

All countries present at the final intergovernmental plenary session held in Johannesburg, South Africa in April 2008 welcome the work of the IAASTD and the uniqueness of this independent multistakeholder and multidisciplinary process, and the scale of the challenge of covering a broad range of complex issues. The Governments present recognize that the Global and Sub-Global Reports are the conclusions of studies by a wide range of scientific authors, experts and development specialists and while presenting an overall consensus on the importance of agricultural knowledge, science and technology for development they also provide a diversity of views on some issues.

All countries see these Reports as a valuable and important contribution to our understanding on agricultural knowledge, science and technology for development recognizing the need to further deepen our understanding of the challenges ahead. This Assessment is a constructive initiative and important contribution that all governments need to take forward to ensure that agricultural knowledge, science and technology fulfils its potential to meet the development and sustainability goals of the reduction of hunger and poverty, the improvement of rural livelihoods and human health, and facilitating equitable, socially, environmentally and economically sustainable development.

In accordance with the above statement, the following governments accept the Synthesis Report.

Armenia, Azerbaijan, Bahrain, Bangladesh, Belize, Benin, Bhutan, Botswana, Brazil, Cameroon, People's Republic of China, Costa Rica, Cuba, Democratic Republic of Congo, Dominican Republic, El Salvador, Ethiopia, Finland, France, Gambia, Ghana, Honduras, India, Iran, Ireland, Kenya, Kyrgyzstan, Lao People's Democratic Republic, Lebanon, Libyan Arab Jamahiriya, Maldives, Republic of Moldova, Mozambique, Namibia, Nigeria, Pakistan, Panama, Paraguay, Philippines, Poland, Republic of Palau, Romania, Saudi Arabia, Senegal, Solomon Islands, Swaziland, Sweden, Switzerland, United Republic of Tanzania, Timor-Leste, Togo, Tunisia, Turkey, Uganda, United Kingdom of Great Britain, Uruguay, Viet Nam, Zambia (58 countries).

While approving the above statement the following governments did not fully accept the Synthesis Report and their reservations are entered in Annex A.

Australia, Canada, United States of America (3 countries).

Part II: Current Conditions, Challenges and Options for Action

Writing team: Inge Armbrecht (Colombia), Nienke Beintema (Netherlands), Rym Ben Zid (Tunisia), Fabrice Dreyfus (France), Shelley Feldman (USA), Ameenah Gurib-Fakim (Mauritius), Hans Hurni (Switzerland), Janice Jiggins (UK), Kawther Latiri (Tunisia), Marianne Lefort (France), Lindela Ndlovu (Zimbabwe), Ivette Perfecto (Puerto Rico), Cristina Plencovich (Argentina), Rajeswari Raina (India), Niels Roling (Netherlands), Elizabeth Robinson (UK), Neils Roling (Netherlands), Hong Yang (Australia)

This assessment of the ways in which knowledge, science and technology contribute to development goals offers a chance to reflect on how people engage their environment to secure healthy lives and livelihoods. Growing concerns with the effects of long-term climatic and ecological changes, which require global as well as national and local responses, make the IAASTD especially opportune. We are, in short, in need of a shared approach to sustainability. This realization is at the heart of the objectives of the IAASTD: how can we reduce hunger and poverty, improve rural livelihoods and facilitate equitable environmentally, socially and economically sustainable development.

This opportunity for stocktaking coincides with the widespread realization that despite significant achievements in our ability to increase agricultural productive capacity to meet growing demand, we have been less attentive to some of the unintended social and ecological consequences of our technological and economic achievements. We are now in a better position to reflect on these costs and to outline policy options to meet the challenges ahead of us, perhaps best characterized as the need for food security under increasingly constrained environmental conditions and globalized economic systems. The IAASTD recognizes the importance of the multiple functions of agriculture and their intersection with other global concerns, including loss of biodiversity and ecosystem services, climate change and water scarcity. Some of the findings from recent assessments conducted by the international community that coincide with those of the IAASTD include:

- Recognition that current social and economic inequities, across and within regions and states, are a significant barrier to achieving development goals.
- Uncertainty about the ability to sustainably produce sufficient food for a continually expanding and demographically changing population where new demands for food and ecosystem services challenge current production systems.
- Uncertainty about the future of world food prices under the impact of climate change, emerging trade regimes, changing dietary patterns and the increased interest in biofuels.
- The end of cheap oil and the need to factor energy efficiency and dependence on tractors, fertilizer, pumped water and transport into food security strategies.
- The emergence of fast-growing economies as additional competitors for resources in the wake of their phenomenal economic growth.
- The increase in chronic ailments, including obesity in poor and rich countries, that increase rates of morbidity and mortality and are partially a consequence of poor nutrition and poor food quality.
- Projected changes in the frequency and severity of extreme weather events in addition to increases in fire hazards, pests and diseases will have significant implications for agricultural production and food security, e.g., for the location of food production, concentrations of human settlements, and water availability.
- The growing awareness of human responsibility for the maintenance of global ecosystem services, and of the changes in global, national and local governance mechanisms required to meet the responsibilities associated with sustainable growth.

We cannot escape our predicament by simply continuing to apply methodological individualism, i.e., by relying on the outcome of individual choices to achieve sustainable and equitable collective outcomes. The IAASTD takes a unique integrated approach to these urgent global problems: the development and deployment of human ingenuity to enhance agriculture, which is defined most broadly to include managing ecological processes in ways that capture and sustain human opportunity. We refer to this as agricultural knowledge, science and technology (AKST). AKST explicitly refers not only to technology but also to the economic and social science knowledge that informs decisions about policies and institutional change required for reaching IAASTD goals. Further, AKST not only refers to "formal" science processes, but also very much to the local and traditional knowledges that still inform most farming today.

IAASTD recognizes that multiple perspectives exist on the nature and role of AKST. For many years, agricultural science focused on delivering component technologies to increase farm-level productivity where the market and institutional arrangements put in place by the state were the primary drivers of the adoption of new technologies. In order to benefit from productivity gains farmers had to continually innovate, reduce farm gate prices and externalize costs.

This model drove the phenomenal achievements of AKST in industrial countries after World War II and the extension of the Green Revolution beginning in the 1960s. But, given the new challenges we confront today, there is increasing recognition within formal S&T organizations that the current AKST model, too, requires adaptation and revision. Business as usual is not an option.

One area of potential adaptation is to move from an exclusive focus on public and private research as the site for R&D toward the democratization of knowledge production. Such an approach requires multiagent involvement to make accessible and available for exchange the skills of local producers. Another area of AKST innovation must lie with more explicit attention to issues that attend to the use of AKST, namely addressing the complex role of institutions, governance practices and social justice concerns that enable or constrain the realization of development and sustainability.

Multifunctionality

The term *multifunctionality* has sometimes been interpreted as having implications for trade and protectionism. This is *not* the definition used here. In IAASTD, multifunctionality is used solely to express the inescapable interconnectedness of agriculture's different roles and functions. The concept of multifunctionality recognizes agriculture as a multi-output activity producing not only commodities (food, feed, fibers, agrofuels, medicinal products and ornamentals), but also non-commodity outputs such as environmental services, landscape amenities and cultural heritages.

The working definition proposed by OECD, which is used by the IAASTD, associates multifunctionality with the particular characteristics of the agricultural production process and its outputs; (1) multiple commodity and non-commodity outputs are jointly produced by agriculture; and (2) some of the non-commodity outputs may exhibit the characteristics of externalities or public goods, such that markets for these goods function poorly or are nonexistent.

The use of the term has been controversial and contested in global trade negotiations, and it has centered on whether "trade-distorting" agricultural subsidies are needed for agriculture to perform its many functions. Proponents argue that current patterns of agricultural subsidies, international trade and related policy frameworks do not stimulate transitions toward equitable agricultural and food trade relation or sustainable food and farming systems and have given rise to perverse impacts on natural resources and agroecologies as well as on human health and nutrition. Opponents argue that attempts to remedy these outcomes by means of trade-related instruments will weaken the efficiency of agricultural trade and lead to further undesirable market distortion; their preferred approach is to address the externalized costs and negative impacts on poverty, the environment, human health and nutrition by other means.

A conception of AKST that includes regulatory frameworks, institutional arrangements, market relations and knowledge in a global economy is reflected in this report. This approach appreciates diverse interests and concerns across a range of agricultural production systems and agricultural producers, including conventional or productivist strategies, agroecological approaches, and indigenous or traditional peasant practices. The IAASTD thus uses the lens of multifunctionality to assess the contribution of AKST to development and sustainability.

In this Report we highlight options drawn from a comparative analysis of the Global and Sub-Global reports (CWANA, ESAP, LAC, NAE and SSA) into two thematic areas: (1) current conditions and major challenges, and (2) options for action.

1. Current Conditions and Challenges

Agriculture and the knowledge systems that are relevant to the sector now face an impasse. There are tremendous achievements in science and production, yet some of the unintended consequences of these very achievements have not been sufficiently addressed. To address these consequences it is important to account for the prevalent inequalities that characterize relations between regions and countries as well as within them. We, as global citizens have little time to lose.

Today we find a world of asymmetric development, unsustainable natural resource use, and continued rural and urban poverty. There is general agreement about the current global environmental and development crisis. It is also known that the consequences of these global changes have the most devastating impacts on the poorest, who historically have had limited entitlements and opportunities for growth.

AKST and agricultural change. Agricultural productivity and production have increased steadily in response to several drivers of change, including the generation and application of AKST. While in North America and Europe (NAE) this phenomenon has been ongoing since the 1940s, in other regions of the world such growth only began in the 1960s, 70s or 80s. In some parts of developing countries formal AKST is yet to make its presence felt as a major driver of agrarian change. The pace of technology generation and adoption has been highly uneven. One region, the NAE, continues to dominate in the volume and variety of agricultural exports, extended value chains and the generation of agricultural technologies (high-yielding varieties, synthetic fertilizers, pesticides and mechanization technologies) as well as recent advances in organic and sustainable production which have helped shape the policies and organizations of AKST in the other regions. While globally, there is an urgent need to revitalize and strengthen AKST, the critical regional differences in agroecosystems, access to formal S&T and diverse impacts on people and ecosystems, pose a challenge to the continuing dominance of a uniform type of formal AKST. The current global system pits small-scale, largely subsistence farmers in rainfed agricultures against farmers who during the past century have been assisted to increasingly capture economies of scale by specialization and externalizing social and environmental costs.

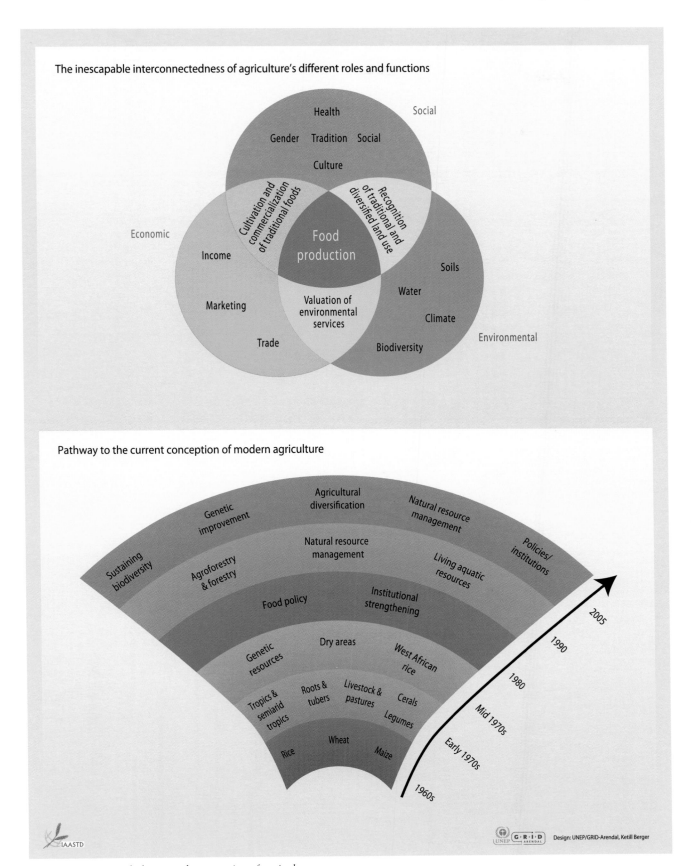

The inescapable interconnectedness of agriculture's different roles and functions

Pathway to the current conception of modern agriculture

Figure SR-P1. *A multifunctional perspective of agriculture.*

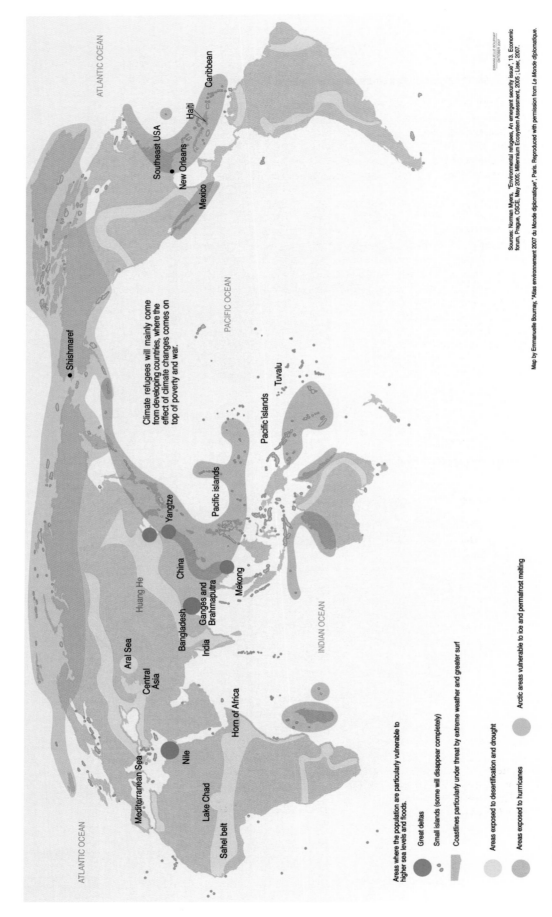

Figure SR-P2. *50 million climate refugees by 2010*

ATLANTIC OCEAN

PACIFIC OCEAN

INDIAN OCEAN

ATLANTIC OCEAN

Caribbean
Haiti
Southeast USA
New Orleans
Mexico

Shishmaref

Climate refugees will mainly come from developing countries, where the effect of climate changes comes on top of poverty and war.

Tuvalu

Pacific islands

Pacific islands

Yangtze
Huang He
China
Mekong

Aral Sea
Central Asia

Bangladesh
Gangès and Brahmaputra
India

Horn of Africa

Mediterranean Sea
Nile
Lake Chad
Sahel belt

Areas where the population are particularly vulnerable to higher sea levels and floods.

- Great deltas
- Small islands (some will disappear completely)
- Coastlines particularly under threat by extreme weather and greater surf
- Areas exposed to desertification and drought
- Areas exposed to hurricanes
- Arctic areas vulnerable to ice and permafrost melting

EMMANUELLE BOURNAY
OCTOBER 2007

Sources: Norman Myers, "Environmental refugees, An emergent security issue", 13. Economic forum, Prague, OSCE, May 2005; Millennium Ecosystem Assessment, 2005 ; Liser, 2007.

Map by Emmanuelle Bournay, "Atlas environnement 2007 du Monde diplomatique", Paris. Reproduced with permission from *Le Monde diplomatique*.

Economic importance, poverty and livelihood expectations. Despite steady growth over the past few decades, the contribution of agriculture to national GDP has been steadily declining in all the regions. The proportion of the population dependent on the sector ranges from 3% in NAE to over 60% in ESAP and SSA. Across diverse geopolitical contexts and ecosystems, agriculture continues to play important economic and social roles and currently engages 2.6 billion people. The majority of the world's poor and hungry live in rural settings and are directly or indirectly dependent on agriculture for their livelihoods.

While the transition from predominantly agrarian economies to industrial or service sector led economies has occurred the world over, the character and rate of industrial growth has been highly differentiated with rural populations surviving on a steadily dwindling share of the economic pie. In addition, agriculture has been subject to worsening terms of trade, globally as well as nationally. The burden of poverty in the sector is incommensurate with the magnitude and range of expectations from agriculture.

AKST and the agricultural and food systems can make a significant contribution to alleviating poverty for the over 1.2 billion people who live on less than $1 per day and provide adequate and nutritious food for the over 800 million undernourished people. Despite a global reduction in absolute poverty, the proportion of the population that is still poor (below poverty line) continues to grow. The need to retool AKST to reduce poverty and provide improved livelihood options for the rural poor—especially landless and peasant communities, urban informal and migrant workers, is a major challenge today. The macro-level challenge is to equip agriculture with the capacity to address the burden of poverty through intra- and inter-sectoral development policies.

Development models and the environment. The drivers of ecological change can best be understood as the consequences of development models pursued over the 20th century. Broadly conceived, the regional imbalance of economic growth, its contribution to the ecological crisis and its effects are differentially experienced in countries of the North and the South. There are multiple causal interlinkages between environmental degradation and poverty, which are exacerbated by the uneven distribution of and access to resources (natural resources, capital, information, etc.) between regions and within countries. For instance, small island nations and the coastal populations of developing countries, which contribute the least to global warming, will be among the first to disappear, yet have very limited if any capacity or resources to respond to such crises.

Across the regions, the poorest, including a disproportionate number of women and children are among the most vulnerable to emerging natural and human-induced environmental disasters. Thus the empowerment of women as repositories of knowledge about local ecosystems, and as significant constituents of the agricultural labor force (62, 66 and 69% in East Asia, SSA and South Asia, respectively) is fundamental to development and to adapting to a changing environment. Parts of CWANA and SSA (e.g., Lesotho, Yemen) still have legislation that denies women land rights and market citizenship.

Even in the well-off countries of NAE where significant knowledge exists about appropriate responses to emerging challenges, actions to address mitigation and adaptation to global climate change have thus far been minimal.

Regional Differences and Achievement of Development and Sustainability Goals

Just as current conditions of agricultural production, environmental degradation, inequality, and availability and access to advanced technologies vary from one region to another, so do the challenges and perception of relative importance of development and sustainability goals. At the global, regional and national levels, decision makers must be acutely conscious of the fact that there are diverse challenges, multiple theoretical frameworks and development models and a wide range of options. Our perception of the challenges and the choices we make at this juncture in history will determine the future of human beings and their environment.

The commitment to address poverty and livelihoods reflects the critical role of agriculture and rural employment opportunities in developing countries where 30-60% of all livelihoods arise from agricultural and allied activities. In NAE, where food insecurity and hunger are no longer major problems, attention has shifted to the question of relative poverty and rapidly declining and changing livelihoods.

Reducing hunger is an important goal in all developing regions: CWANA, ESAP, LAC and SSA. Of the 854 million malnourished people in 2001 to 2003, only 9 million were in the developed world; ESAP accounted for 61% of the total. In ESAP, however, this represents only 15% of the total regional population while the 206 million malnourished SSA inhabitants represent 32% of the region's population. The substantial number of hungry and malnourished people in NAE indicates that more production does not necessarily equate with hunger reduction.

Improving human health and nutrition is critical for all regions. AKST can affect health via food safety and security, chronic and infectious diseases, and occupational health. Malnutrition is a major cause of ill health and reduced productivity, particularly in SSA and CWANA. Food safety is an important health issue in all regions. Inappropriate application of AKST contributes to the increase in overweight, obesity, and chronic diseases that is being experienced in all countries. The burden of emerging and reemerging infectious diseases remains high in SSA, CWANA, and ESAP. The relative burden of occupational health burdens is lowest in NAE.

Environmental goals are important globally despite pressure on the environment due to relatively high industrialization, urbanization and productivity enhancing agricultural practices in NAE, and pressures to enhance productivity even at the cost of environmental goods and services in SSA. This is consistent with the relative contribution of agriculture to natural resource degradation, as well as to the relative importance of agriculture in the overall economy in each region, as is evident in their respective IAASTD Summaries for Decision Makers.

Equity is important across all regions. This goal draws attention to the current conditions of iniquitous distribution and access to resources and to overall income inequality, which is most extreme in LAC. Regional analyses (ESAP,

LAC and SSA) indicate that the unequal distribution of resources is a major constraint that shapes development needs and impedes the achievement of all other development and sustainability goals.

Farming systems

Agriculture is currently constrained in its capacity to respond to poverty and generate a range of livelihood options in rural areas. Farming systems are very diverse and range between large scale capital intensive farming systems to small-scale labor intensive farming systems. Over the 20th century there was increasing farming system specialization in NAE, largely due to the implementation of policies and measures aimed at expanding agricultural production (land reclamation, subsidies, price systems, border tariffs). A high proportion of farmers in CWANA, ESAP, LAC and SSA are small-scale producers whose livelihood strategies include poly-cropping, tree products and livestock as well as off-farm activities. In developing countries generally, limited rural and urban employment opportunities and the continuing dependence of cultivators on economically unviable small-scale holdings (increasing input prices; relatively stagnant agricultural output prices; cheap, subsidized imports; and limited surplus) have diminished the viability of subsistence production alone.

In addition, modern biological, chemical and mechanical technologies, in particular, are designed for farms and farming systems which have attendant entitlements and conditions that enable the production of tradable and vertically integrated commodities in value chains. Where the government and some private and civil society organizations have enabled appropriate scale effects as well as technical and financial support, small-scale farmers also have intensified their production systems and benefited from increasing market integration. Though the productivity per unit of land and per unit of energy use is much higher in these small and diversified farms than the large intensive farming systems in irrigated areas, they continue to be neglected by formal AKST. [See Part II: Bioenergy and Climate Change]

In the semiarid CWANA where water scarcity is prevalent, current conditions favor large-scale monocropping systems that rely on high investment (in water supply, machinery and agrochemicals) and cause environmental degradation, although positive solutions can emerge through AKST and incentives for enhancing incomes in the small-scale farm sector. The challenge for AKST is to address these small-scale farms in diverse ecosystems and to create realistic opportunities for their development; the potential for improved area productivity is decreasing, except for low-input and labor-oriented agriculture in a few regions of the world.

There is a significant correlation between capital stock in agriculture and value added per worker—for example in CWANA, countries with capital intensive agriculture are associated with high value added per worker. In many developing countries, especially in SSA and the least developed countries in ESAP, the low capitalization of agriculture translates into low value added per worker, thus worsening the vicious cycle of agrarian and rural poverty. These conditions are often coupled with declining employment opportunities in agriculture that require rural laborers to secure alternative non-farm employment. Unfortunately,

the non-farm labor market is constrained by high unemployment, especially for the relatively large unskilled young population in search of work. While organic and ecological agriculture as practiced in parts of ESAP and LAC can provide more employment, the absolute unemployment figures, especially in ESAP, are massive. In SSA and ESAP as well as labor surplus countries in other regions, it is crucial to explore how agricultural and rural production processes can be better linked with industrial and service sector growth. AKST in its current form, whether as formal S&T organizations or local and traditional knowledge specific to agro-ecosystems, is limited in its capacity to inform change in the institutions that frame human interaction, equitable and just governance and vibrant links with other sectors of the economy.

Market conditions, trends and challenges

Agricultural commodities the world over are currently facing a secular decline in prices accompanied by wide fluctuations. IAASTD projections of the global food system indicate a tightening of world food markets, with increasing market concentration in a few hands and rapid growth of global retail chains in all developing countries, natural and physical resource scarcity, and adverse implications for food security. Real world prices of most cereals and meats are projected to increase in the coming decades, dramatically reversing past trends. Millions of small-scale producers and landless labor in developing countries and underdeveloped markets, already weakened by changes in global and regional trade, with poor market infrastructure, inadequate bargaining capacity and lack of skills to comply with new market demands, will face reduced access to food and livelihoods.

The food security challenge is likely to worsen if markets and market-driven agricultural production systems continue to grow in a "business as usual" mode. By 2050, the world will have 80 million severely malnourished children, concentrated mainly in South Asia and sub-Saharan Africa. Industrialized country agricultural subsidies and advantages in agricultural added value per worker close off options for the export of agricultural commodities from sub-Saharan Africa and distort their domestic markets, thereby suppressing producer incentives to adopt new technologies and enhance crop productivity. In CWANA and ESAP, trade barriers (including IPR, quality standards), market distorting domestic policies and international protocols or restrictions add to the complexity of future food security. The food security challenge is likely to worsen current conflicts, cross border tensions, and environmental security concerns.

In CWANA, ESAP, LAC, and SSA, a number of mechanisms to protect producers from price fluctuations and enable access to and compliance with new market practices or trade requirements (like sanitary and phytosanitary [SPS] measures), include market-based instruments such as futures trading, which small-scale producers find difficult to access. Market based instruments also include commodity boards and price regulation which large buyers find too limiting to meet their needs [See Part II: Trade and Markets]. The emergence of regional and preferential trade agreements and trading blocs among developing countries reveals an increasing mistrust of, and untenable nature of global trade regimes, given the perception of an unequal playing field.

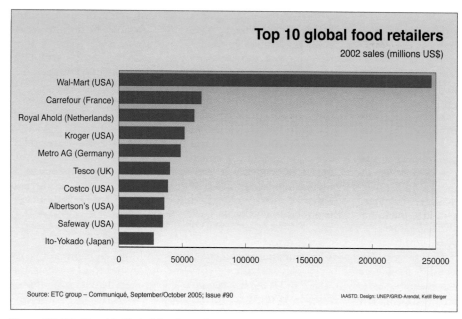

Figure SR-P3. *Top 10 Global food retailers.*

However, overall, given the complex socioeconomic contexts, geopolitical and ecological processes in the agricultural and allied sectors, markets tempered with appropriate state support and regulation can be effective instruments to address poverty, livelihood needs and income, as well as environmental services and responsibilities of agriculture.

Multifunctional agricultural systems

By definition, the principle of multifunctionality in agriculture refers to agriculture that provides food products for consumers, livelihoods and incomes for producers and a range of public and private goods and services for citizens and the environment, including ecosystem functions. Existing specialization in the global agrifood system, coupled with government investments and policies in production and trade has led to a view of agriculture as an exclusively economic activity, measured in commodity-based, monetary terms. In the specialized production systems of NAE and parts of ESAP, CWANA and LAC, the focus on the multiple roles and functions of agriculture is drawing policy attention largely in response to the scope of possible investments in indirect support mechanisms, production and trade. In the relatively less endowed and more diverse farming systems of the world, especially in SSA and large parts of LAC, ESAP, and CWANA, the multiple functions of agriculture are being addressed as an important way to reduce the loss of biodiversity, encourage ecofriendly production systems and local and traditional knowledge, improve nutrition and gender relationships in agriculture through diverse production and processing systems and maintain a suite of livelihood options in rural areas.

These region-specific agricultural systems have the potential to be either highly vulnerable or sustainable, due to the inescapable interconnectedness and tradeoffs between the different roles and functions of agriculture. Formal AKST has typically focused on increased specialization of commodity production and not on optimizing the outcomes from dynamically evolving multifunctional systems involving biophysical and socioeconomic components. A challenge that AKST needs to overcome is the lack of research in geographical, social, ecological, anthropological and other evolutionary sciences as applied to diverse agricultural ecosystems. These are necessary to devise, improve and create management options and contribute to multifunctionality and may help in improving the sustainability of these resources and their effective use in production systems.

The social and cultural implications of livelihood options and of poverty, nutrition, and ecosystem conservation, whether of highly productive mixed crop-livestock systems in the wetlands or of low productivity crop-fodder-fiber and small ruminants systems in the arid areas in SSA, differ from the sociocultural implications of livelihoods and incomes from commercial production in France and California. Similarly, current subsidies, tariffs and investments to agriculture in countries like India, China and Japan in ESAP, and Tunisia and Syria in CWANA, imply different conditions, interests and capacities to address the tradeoff between the production and environmental functions of agriculture. As learned from the much contested sugar and cotton production and trade disputes, relative economic and environmental vulnerability, differential state support, agribusiness systems and market regulations determine the interconnectedness of the economic, social and environmental functions of agriculture. There is increasing recognition of the multiple roles and functions of agriculture, which can address environmental sustainability, poverty reduction and help achieve the elimination of hunger and malnutrition. The main challenges posed by multifunctional agricultural systems for AKST are:

• How do we support the necessary tradeoffs among increasing the productivity of food and animal feed to meet changing food habits, and enabling fiber and fuel wood production, while satisfying increasing current and emerging energy demands, as well as environmental and cultural services by agroecosystems?

- How do we practically provide clean water, maintain biodiversity, sustain the natural resource base and decrease the adverse impacts of agricultural activities on people and the environment?
- How do we improve social welfare and personal livelihoods in the agricultural sector, and enhance these economic benefits for the other sectors?
- How do we empower marginalized stakeholders to sustain the diversity of agriculture and food systems, including their cultural dimensions?
- And how do we increase productivity under marginalized, rainfed lands and incorporate them into local, national and global markets?

Resource use and degradation

Changes in land use have been without exception significant in all the regions. While more land has been brought under the plough in SSA over the past two decades than during any period of human history on the sub-continent, the intensification of production without the expansion of land under cultivation has been significant in NAE, ESAP and LAC. In much of CWANA, such expansion is constrained by access to water. Agriculture has contributed to land degradation in all the regions; in some regions with input intensive production systems (ESAP, LAC and NAE) the relative share of agriculture-induced degradation is higher than in other regions. On average 35% of severely degraded land worldwide is due to agricultural activities.

Poorly defined and enforced property rights over common pool resources (SSA), lack of property rights for women (CWANA, ESAP, LAC, SSA), and caste and other social hierarchies that limit access to resources (ESAP, LAC, SSA) have contributed natural resource degradation. Overall population growth, increasing pressure to generate income from natural resources (using increasingly expensive inputs), and technological solutions that are blanket recommendations irrespective of regional variations in resource quality, have intensified production and extraction processes of crop/commodity production, livestock, fisheries and forestry. As a result, pockets of high-input agriculture in CWANA, ESAP and LAC as well as the NAE region contribute to the degradation of soil and water systems and pollution that add to global warming. These conditions confront limited state capacities to cope with the effects of climate change in the developing countries [See Part 2: NRM and Climate Change].

The complex nexus between degradation of natural resources and rural poverty is acknowledged in the drylands of SSA, South Asia and CWANA, mountain ecosystems of LAC and coastal ecosystems in all the regions. Despite evidence of several resource conserving technologies and resource sharing and improving social contracts or institutional arrangements, little effort has been made within mainstream formal AKST to learn from and apply these lessons to other agroecological systems and societies. Moreover, while declining water availability and quality, the loss of biodiversity, farmer access to seeds and local plant and animal genetic resources, and local capacities to mitigate and adapt to climate change are discussed in the regions, little effort has thus far been

made to address the causal factors (such as lack of assured property rights and tenure laws, absence of incentives for conservation, and subsidies to address resource constraints) that support resource exploitative production. Environmental technologies such as integrated pest management, agroforestry, low-input agriculture, conservation tillage, pest resistant GM crops, and climate change adaptations, have often faced a policy gridlock with formal AKST, civil society, the state, private industry and media taking highly polarized positions. Now as biofuels and plantation agriculture add to the competition for limited natural resources, the tradeoffs between production and environmental benefits must be increasingly scrutinized. The challenge is to maintain and enhance environmental quality for increased agricultural production and other goods and services.

Social equity

Worsening income inequality is a serious concern and poses a significant challenge for agricultural and food systems and AKST in all the five regions. The uneven distribution of productive natural resources coupled with the lack of access to resources and fair markets for small-scale producers and women in agriculture, results in extreme inequality and increasing poverty. While peasants and women cultivators are uncommon in NAE, millions of poor people and women in much of CWANA, ESAP, LAC, and SSA contend with unequal production and market relationships on a daily basis. Current inequality is exacerbated by the fact that NAE dominates agricultural and rural development resources as well as formal knowledge generation in AKST. For example, businesses within NAE have a powerful impact on global consumer demand; they obtain and profit, directly or indirectly, from commodities, landraces and other valuable genetic resources (stored *ex situ* in other countries), beneficial organisms for biocontrol programs, immigrant labor and have legal and institutional capacities such as intellectual property rights, standards and market regulations, which many countries in the developing regions lack.

Landless agricultural labor is at the receiving end of inequitable distribution of productive resources, production practices and technologies. There is increasing rural to urban male migration in search of employment in all developing countries. Social security nets and the provision of non-farm rural or urban employment opportunities are being attempted by countries along with proactive local employment and income generation programs spearheaded by the CSOs. However, these programs remain limited in both scale and scope.

All five regions are acutely conscious of increasing indigence and social exclusion of several indigenous and tribal peoples. Many of these communities are repositories of traditional knowledge and fast depleting, but highly valuable knowledge about local ecosystems and processes of change and management. Much of this knowledge is outside the purview of modern AKST and is increasingly subject to pressure from commercial crop, livestock, fisheries or forest-based production [See Part II: Traditional and Local Knowledge]. Within formal AKST systems, little has been done to acknowledge or address the livelihoods concerns, technological and development needs of women, labor and indigenous peoples. Instead, over the past several decades,

AKST and current agricultural development models have contributed to increasing inequality and the exclusion of indigenous and tribal peoples.

In LAC and parts of ESAP the selective perception of production requirements and exclusion of or limited attention given to certain agroecosystems, such as dryland agriculture, coastal fisheries, mountain ecosystems, and pastoral systems, worsens the inequality already compounded by local exploitation, rent seeking and corruption, appropriation of resources of the poor—especially common pool resources—and social prejudices like caste and gender biases. The challenge for development policy and AKST is to develop agricultural and food systems that can reduce income inequalities and ensure fair access to production inputs and knowledge to all. Governments and international donors are now beginning to invest in long-term commitments to AKST integrated into pro-poor development policies.

AKST—Current constraints, challenges and opportunities

More than five decades after formal AKST made its entry into almost all countries, the explicit economic and political legitimization of investments in AKST remains food security, livelihoods and poverty reduction in developing countries, and trade and environmental sustainability in industrialized countries. While the development models-poverty-environmental degradation nexus is evident in different forms in different countries, the formal AKST apparatus available to address these variations is the same in structure, content and the conduct of science in almost all countries. The AKST apparatus tends to focus on mainstream, input-intensive, irrigated monocropping systems—mainly cereals, livestock and other trade-oriented commodities, to the relative neglect of arid/dryland agriculture, mountain ecosystems, and other non-mainstream production systems that have been discussed above. It is important to recognize that this constraint, more or less universal in formal AKST is not incidental, but part of an overall development model in which scientific knowledge is institutionalized in its utilitarian role. Resources are allocated to production systems that can show the highest economic returns to crop/commodity productivity. The capacity of AKST to address the challenges of poverty, livelihoods, health and nutrition, and environmental quality is conditioned by its capacity to address its own internal constraints and challenges.

Organized AKST in the form of public sector R&D, extension and agricultural education across world regions, are based upon a linear top-down flow of technologies and information from scientific research to adopters. Despite increasing polarization of the debate on new technologies, especially biotechnology and transgenics, and years of well-published knowledge on differential access to technologies and appropriate institutional arrangements, formal AKST has yet to address the question of democratic technology choice. AKST as currently organized in public and private sector does little to interact with academic initiatives in basic biological, ecological and social sciences to design rules, norms and legal systems for market-oriented innovation and demand-led technology generation, access and use appropriate for meeting development and sustainability goals.

There is a significant volume of literature from all the regions on the high rates of return per unit of investment in agricultural R&D, especially in crops and in farming systems that have been the focus of the AKST apparatus. Some of the conditioning factors for high rates of return lie outside agriculture and AKST in complementary investments such as rural infrastructure or microcredit units that reduce market transaction costs or provide appropriate institutions or norms. A rate of return analysis is insufficient for capturing returns to investment that meet development and sustainability goals; other economic and social science methods are needed for this task.

Declining investments in formal AKST by international donors and a number of national governments is causing concern among the developed and developing countries. Public investments in agricultural R&D continue to grow although rates have declined during the 1990s. In many industrialized countries investment has stalled or declined, while in ESAP countries investments have grown relative to other regions (annual growth rate of 3.9% in the 1990s). As a result, ESAP accounts for an increasing share of global public R&D investment, from 20% in 1981 to 33% in 2000. In contrast to the 1980s, the annual growth rate of total spending in SSA decreased in the 1990s from 1.3 to 0.8%. A disturbing trend in 26 SSA countries for which time series data are available is that the public sector spent less on agricultural R&D in 2000 than a decade earlier. Globally public sector R&D is becoming increasingly concentrated in a handful of countries. Among the rich countries, just two, the USA and Japan, accounted for 54% of public spending in 2000, and three developing countries, China, India and Brazil, accounted for 47% of the developing world's public agricultural research expenditures. Meanwhile, only 6% of the agricultural R&D investments worldwide were spent in 80 mostly low-income countries whose combined population in 2000 was more than 600 million people.

In the industrialized countries investment by the private sector has increased and is now higher than total public sector investments. In contrast, private sector investment in developing countries is small and will likely remain so given weak funding incentives for private research. In 2000, private firms invested only 6% of total spending in the developing world, of which more than half was invested in ESAP. Private investment in AKST is, and is likely to remain, largely confined to appropriable technologies, with intellectual property protection, which can earn significant revenues in the market.

Currently AKST actors and organizations are not sufficiently able to deal with the challenges ahead because of the focus on too narrow a set of output goals. The current knowledge infrastructure, which is oriented toward these goals, historically has largely excluded ecological, environmental, local and traditional knowledges and the social sciences. AKST infrastructure will need to encompass and work with this much broader set of understanding and data if AKST challenges are to be met. The knowledge infrastructure of AKST is closely allied with particular branches of economics appropriate for meeting production goals, but to the relative neglect of other capacities in the economic sciences that are needed to meet AKST challenges.

Meeting the challenges will require a different organizational framework than currently exists in fundamental

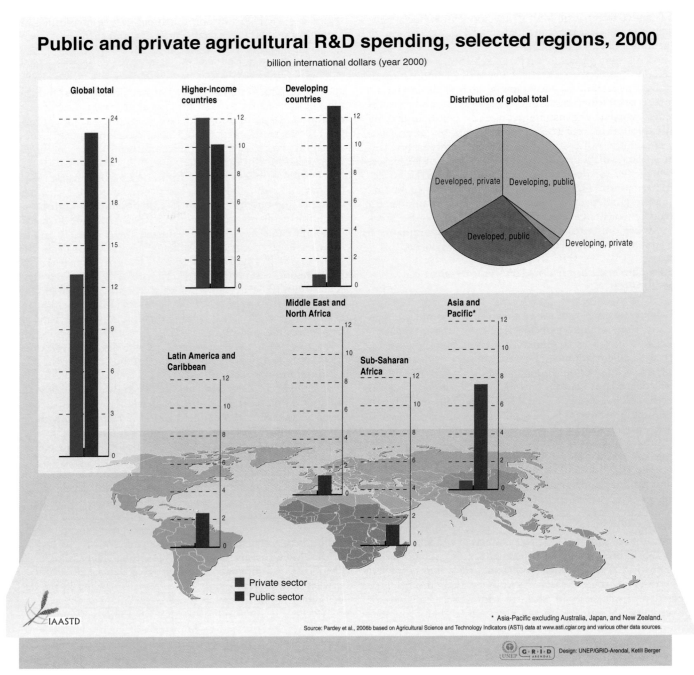

Public and private agricultural R&D spending, selected regions, 2000

billion international dollars (year 2000)

Global total

Higher-income countries

Developing countries

Distribution of global total

Developed, private | Developing, public

Developed, public

Developing, private

Latin America and Caribbean

Middle East and North Africa

Sub-Saharan Africa

Asia and Pacific*

■ Private sector
■ Public sector

IAASTD

* Asia-Pacific excluding Australia, Japan, and New Zealand.

Source: Pardey et al., 2006b based on Agricultural Science and Technology Indicators (ASTI) data at www.asti.cgiar.org and various other data sources.

UNEP G·R·I·D ARENDAL Design: UNEP/GRID-Arendal, Ketill Berger

Figure SR-P4. *Public and Private Agricultural R&D Spending by Region, 2000.*

and applied scientific capability. Breakthroughs in advance science will not lead to relevant effective and efficient applications that address development and sustainability unless investments in public, commercial and civil society at local levels are sustained or increased. The challenges ahead demand a greater focus on management systems—from crop to whole farm to natural resource area, landscape, river system and catchment scales. Management systems require sophisticated understanding of the institutional dimensions of management practices and of decision processes that must

be coordinated across variable spatial, temporal and hierarchical scales. AKST specialists will need a more profound understanding of the legal and policy frameworks that increasingly will steer agricultural and food system development.

Emerging challenges. In all the regions, there is an overarching concern with poverty and livelihoods among the relatively poor, which are faced with intra- and inter-regional inequalities. The willingness of different actors, including those in the state, civil society and private sector, to address the fundamental question of the relationships among production, social and environmental systems is marred by contentious political and economic stances adopted by the different actors. The acknowledgment of current challenges and the acceptance of options available for action require a long-term commitment from decision makers that is responsive to specific needs and a wide range of stakeholders. It calls for a continuing recognition that science, technology, knowledge systems and human ingenuity are needed to meet future challenges, opportunities and uncertainties.

2. Options for Action
Successfully meeting development and sustainability goals and responding to new priorities and changing circumstances will require a fundamental shift in science and technologies, policies and institutions, as well as capacity development and investments. Such a shift will recognize and give increased importance to the multifunctionality of agriculture and account for the complexity of agricultural systems within diverse social and ecological contexts. Successfully making this shift will depend on adapting and reforming existing institutional and organizational arrangements and on further institutional and organizational development to promote an integrated approach to AKST development and deployment. It will further require increased public investment in AKST and development of supporting policy regimes.

Poverty and livelihoods
Ensuring the development, adaptation and utilization of formal AKST by small-scale farmers requires acknowledging the inherently diverse conditions in which they live and work. Hence, formal AKST needs to be informed by knowledge about farmers' conditions, opportunities and needs, and by participatory methodologies that can empower small-scale producers. The development of more sustainable low-input practices to improve soil, nutrient and water management will be particularly critical for communities with limited access to markets. Enabling resource-poor farmers to link their own local knowledge to external expert and scientific knowledge for innovative management of soil fertility, crop genetic diversity, and natural resources is a powerful tool for enabling them to capture market opportunities

Technological innovation at the farm level is predicated upon enabling institutional and legal frameworks and support structures, such as:
* Giving producers a voice in the procedures for funding, designing and executing formal AKST;
* Enhancing producer livelihoods though brokered long-term contractual arrangements, through commercial out-grower schemes or farmer cooperatives. They involve commodity chains that integrate microcredit, farmer organization, input provision, quality control, storage, bulking, packaging, transport, etc.;
* Investments to generate sustainable employment opportunities for the rural poor, both landless labor and cultivator households, e.g., through enhanced value-added activity and off-farm employment;
* Promoting innovation grounded in interaction among stakeholders who hold complementary parts of the so-

lution, e.g., farmers, technical specialists, local government agents, and private input traders.

Though these interactions take place at the decentralized level, they usually require enabling conditions at higher levels that include legal frameworks that ensure access and secure tenure to resources and land, recourse to fair conflict resolution and other mechanisms for accountability and national policies that support remunerative farm prices.

Policy options to increase domestic farm gate prices for small-scale producers include:
* Fiscal policy (e.g., market feeder roads, postharvest storage facilities and rural value-added agrifood production) to develop infrastructural capacity, and increasing the percentage of that small-scale farmers receive for export crops;
* Acknowledgment of access to (market and policy) information, farmer-to-farmer exchange, farmer education, and extension as public service and public goods that provide access to AKST both formal and local. In LAC, for example, farmer-to-farmer approaches have proven successful in the adoption of agroecological practices;
* Public/private arrangements that allow producers to sell through urban supermarkets;
* Strengthening producer organizations through investment in travel and meetings, and capacity building and through creating space for farmer participation in local, regional and national decision making; and
* Capturing preferential trading arrangements.

Farmer Field Schools, Participatory Plant Breeding/Domestication, Farmer Research Groups and similar forms of interaction in support of farmer-driven agendas have been shown to have multiple pro-poor benefits, such as enduring farmer education, empowerment and organizational skills [see Part II: NRM].

Developments are needed that build trust and that value farmer knowledge, agricultural and natural biodiversity, farmer-managed medicinal plants, local seed systems and common pool resource management regimes. The success of options implemented locally rests on regional and nationally based mechanisms to ensure accountability.

Food security
Food security is a situation that exists when all people, at all times, have physical, social and economic access to sufficient, safe and nutritious food that meets their dietary needs and food preferences for an active and healthy life. *Food sovereignty* is defined as the right of peoples and sovereign states to democratically determine their own agricultural and food policies.

Using appropriate AKST can contribute to radically improved food security. It can support efforts to increase production, enhance the social and economic performance of agricultural systems as a basis for sustainable rural and community livelihoods, rehabilitate degraded land, and reduce environmental and health risks associated with food production and consumption. The following options can aid in capturing these opportunities to increase sustainable agricultural production:

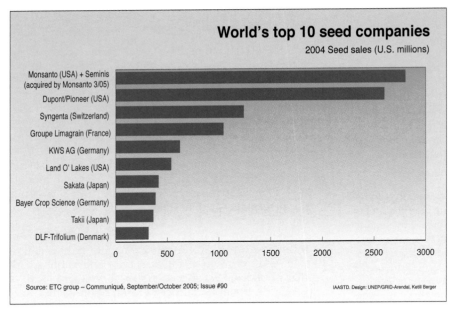

World's top 10 seed companies
2004 Seed sales (U.S. millions)

Source: ETC group – Communiqué, September/October 2005; Issue #90

IAASTD. Design: UNEP/GRID-Arendal, Ketill Berger

Figure SR-P5. *Global vegetable seed market shares.*

- Expanding use of local and formal AKST (e.g., conventional breeding, participatory decentralized breeding and biotechnology) to develop and deploy suitable cultivars (millets, pulses, oilseeds, etc.) and better agronomic practices that can be adapted to site-specific conditions [CWANA; ESAP; SSA].
- Breeding and improvement work on some minor crops in different subregions.
- Improving soil, water and nutrient management and conservation of biodiversity [CWANA; ESAP; LAC; SSA; SR Part II: NRM] and improving access to resources (e.g., nutrients and water) [SSA].
- Increasing small-scale diversification by enhancing the role of animal production systems, aquaculture, agroforestry with indigenous fruits and nuts, and insects [CWANA; ESAP; SSA; Part II: NRM].
- Enabling an evaluation culture within AKST with appropriate incentives to assess the past and potential impacts of technological and institutional changes deployed in the field.

Important to consider when shifting from food crops to biofuels on the basis of economic feasibility is attention to the impact of large areas devoted to such crops on food security and the environment [ESAP, LAC, SSA; SR Part II: Bioenergy].

Some of the AKST policy options for addressing food security include:
- Mobilizing the productive capacity and sustainability of rain fed areas;
- Addressing price fluctuations and reductions through market instruments that enable shifting risk away from vulnerable small-scale producers;
- Reducing transaction costs and creating special access rights in regional and global trade for millions of small-scale producers; social security nets for women and highly vulnerable indigenous and tribal populations to ensure access to affordable and safe food;
- Strengthening local markets by improving the connection between rural areas and cities; food producers and urban food consumers; and urban and peri-urban agriculture producers and consumers [LAC]; and
- Improving food safety and quality through the enforcement of enhanced regulatory and monitoring regimes.

Public sector research has yet to offer a range of viable rural management and agronomic practices for crop and livestock systems that are appropriate for water-restrained dry lands and poor farmers [CWANA; ESAP; SSA]. Private sector research, concentrated on internationally traded crops, is less likely to find such projects profitable, at least in the immediate future. Yet, public funding for such research in these crops and regions will be necessary if we are to address the needed changes in organizational and institutional arrangements to respond to the constraints imposed by poor management systems. Such investments will likely assist in limiting natural resource degradation and environmental deterioration, and contribute to decreasing the poverty and pockets of hunger that currently persist in the midst of prosperity [ESAP].

Environment
- *Knowledge, science and technology (local and formal).* "Business as usual" is not an option if we want to achieve environmental sustainability. To help realize this goal, AKST systems must enhance sustainability while maintaining productivity in ways that protect the natural resource base and ecological provisioning of agricultural systems. Options include: Improving energy, water and land use efficiency through the use of local and formal knowledge to develop and adapt site-specific technologies that can help maintain, create or restore soils, increase water use efficiency and reduce

contamination from agrochemicals [CWANA; ESAP; Global Chapter 3; LAC; SSA; SR Part II: NRM].

- Improving the understanding of soil-plant-water dynamics, that is, ecological processes in soil and bodies of water and ecological interactions that affect agricultural and other natural resources systems [Global Chapter 3; LAC; NAE].
- Creating and improving management options to support agroecological systems (including landscape mosaics) and the multiple roles and functions of agriculture with input from ecological and evolutionary science practitioners, plant geneticists, botanists, molecular biologists, etc. [Global Chapter 3; SR Part II: NRM].
- Increasing our knowledge of local and traditional knowledge to support learning more about options for sustainable land management and rehabilitation [Global Chapter 3; Part II: NRM].
- Enhancing *in situ* and *ex situ* conservation of agrobiodiversity through broad participatory efforts to conserve germplasm and recapture the diversity of plant and animal species traditionally used by local and indigenous people [Global Chapter 3; LAC; NAE; SSA; SR Part II: NRM]. Strengthening plant and livestock breeding programs to adapt to emerging demands, local conditions, and climate change [SSA]. Increasing knowledge and providing guidelines for the sustainable management of forest and fisheries and integrating them within farming systems in such a way to maximize the income and employment generation in rural areas [Global Chapter 3; SR Part II: NRM]. Democratically evaluating existing and emerging technologies, such as transgenic crops, first and second generation biofuels, and nanotechnologies to ascertain their environmental, health and social impacts [Global Chapter 3; LAC; NAE]. Long-term assessments are needed for technologies that require considerable financial investment and risk to adopters, such as biotechnology and Green Revolution-type technologies (high external inputs). It is important that impacts and applications of alternative technologies are also examined and that independent comparative assessments (i.e., comparing transgenic with currently available agroecological approaches such as biological control) are conducted. Improving the understanding of the agroecological functioning of mosaics of crop production areas and natural habitats, to determine how these can be co-managed to reduce conflicts and enhance positive synergies. Promoting more diverse systems of local crop production at farm and landscape scale, to create more diverse habitats for wild species/ecological communities and for the provision of ecosystem services. This will require institutional innovations to enable efficient marketing systems to handle diversified production. Establishing decentralized, locally based, highly efficient energy systems and energy efficient agriculture to improve livelihoods and reduce carbon emissions [ESAP; LAC]. AKST can contribute to the development of economically feasible biofuels and biomaterials that have a positive energy and environmental balance and that will not compromise the world food supply [Global Chapter 3; NAE; SR Part II: Bioenergy, NRM]. Developing strategies to counter the effects of agricul-

ture on climate change and strategies to mitigate the negative impacts of climate change on agriculture [Global Chapter 3; SR Part II: NRM].

Reducing agricultural emissions of greenhouse gases will require changes to farming and livestock systems and practices throughout the food system [NAE; LAC] as well as land use changes to achieve net carbon sequestration. Better agronomic practices, especially in livestock and rice production, such as conservation agriculture, less water consuming cultivation methods, and improved rangeland management, feeding of ruminants and manure management, can substantially reduce GHG emissions and possibly increase C sequestration [CWANA; ESAP]. Agroecological methods, agroforestry, and the breeding of salt-tolerant varieties can help mitigate the impacts of climate change on agriculture [ESAP; LAC; SSA; SR Part II: Climate change]. Although knowledge in these areas already exists, it is important to analyze why this knowledge is not applied more often.

Policies and institutional frameworks. Options need to reflect the goals of sustainable development and the multiple functions of agriculture, being particularly attentive to the interface between institutions and the adoption of AKST and its impacts. To be effective in terms of development and sustainability, these policies and institutional changes should be directed primarily at those who have been served least by previous AKST approaches, i.e., resource-poor farmers.

Policies that promote sustainable agricultural practices, e.g., using market and other mechanisms to regulate and generate rewards for agro/environmental services, stimulate more rapid adoption of AKST for better natural resource management and enhanced environmental quality should be considered to promote more sustainable development [Global]. Some examples of sustainable initiatives are policies designed to:

- Reduce agrochemical inputs (particularly pesticides and synthetic fertilizers);
- Use energy, water and land more efficiently (not only as in precision agriculture, but also as in agroecology);
- Diversify agricultural systems;
- Use agroecological management approaches; and
- Coordinate biodiversity and ecosystem service management policies with agricultural policies [CWANA; ESAP; Global Chapter 3; LAC].
- Internalize the environmental cost of unsustainable practices [ESAP; Global Chapter 3; LAC; NAE] and avoid those that promote the wasteful use of inputs (pesticides and fertilizers);
- Ensure the fair compensation of ecosystem services [CWANA; ESAP; Global; LAC; NAE; SSA];
- Regulate environmentally damaging practices and develop capacities for institutional changes that ensure monitoring and evaluation of compliance mechanisms [ESAP; Global].
- Facilitate and provide incentives for alternative markets such as green products, certification for sustainable forest and fisheries practices and organic agriculture [CWANA; ESAP; Global; LAC; NAE; SSA] and the strengthening of local markets including enhancing

intra-region links between rural producers and urban consumers [LAC];

- Enable resource resource-poor farmers to use their traditional and local technical knowledge to manage soil fertility, crop and livestock genetic diversity and conserve natural resource (e.g., microcredit for transitioning toward agroecological practices, processing, and production) to make them sustainable and economically viable;
- Adjust intellectual property rights (IPR) and related framework to allow farmers to managed their seeds and germplasm resources as they wish.

To achieve more sustainable management, institutional and socioeconomic measures are required for the widespread adoption of sustainable practices. Long-term land and water use rights (e.g., land and tree tenure), risk reduction measures (safety nets, credit, insurance, etc.) and establishing profitability of recommended technologies are prerequisites for adoption. For resources with common pool characteristics, common property regimes are needed that most likely will be developed by rural communities and supported by appropriate state institutions. Farmers need guaranteed long-term access to the resources necessary for the implementation of culturally and technically appropriate sustainable practices [Global Chapter 3]. Also needed are new modes of governance that emphasize participatory and democratic approaches and require the development of innovative local networks. Institutional reforms, too, are needed to enable formal AKST to partner effectively with small-scale producers, women, pastoralists, and indigenous and tribal peoples who are sources of environmental knowledge. Stakeholder monitoring of environmental quality can help develop production technologies and environmental services [ESAP; Global Chapter 3].

Given existing and increasing conflicts over natural resources and environmental insecurity (e.g., disputes over fishing rights, water sharing, climate change mitigation), policies, agreements and treaties that promote regional and international cooperation can assist in realizing the development and sustainability goals. Conflict resolution systems for managing conservation programs, monitoring pest and disease incidence, and monitoring development and compliance mechanisms would also help in realizing these goals [ESAP; Global Chapter 3].

There is significant scope for AKST and supporting policies to contribute to more sustainable fisheries and aquaculture that can contribute to reducing overfishing. Yet many governments still struggle to translate guidelines and policies into effective interventions able to provide an ecosystem approach to fisheries management. At the least policies are needed to end subsidies that encourage unsustainable practices (e.g., bottom trawling). Small-scale fisheries need explicit support and the promotion of increased awareness of sustainable fishing practices and postharvest technologies, as well as policies that reduce industrial scale fishing. Implications of increased aquaculture production (e.g., loss of coastal habitats, increased antibiotic use, etc.), and catch fisheries should also be considered.

Regardless of the differing opinions about transgenics in the regions, all Sub-Global reports recognized the importance of assessing both the potential environmental, health and social impacts of any new technology, and the appropriate implementation of regulatory frameworks as a principled matter of precaution. Particular concerns exist regarding potential genetic contamination in centers of origin [See SR Part II: Biotechnology].

The formal AKST system is not well equipped to promote the transition toward sustainability. Current ways of organizing technology generation and diffusion will be increasingly inadequate to address emerging environmental challenges, the multifunctionality of agriculture, the loss of biodiversity, and climate change. Focusing AKST systems and actors on sustainability requires a new approach and worldview to guide the development of knowledge, science and technology as well as the policies and institutional changes to enable their sustainability. It also requires a new approach in the knowledge base; the following are important options:

- The revalorization of traditional and local knowledge [CWANA; ESAP; Global; LAC; NAE; SSA] and their interaction with formal science;
- An interdisciplinary (social, biophysical, political and legal), holistic and system-based approaches to knowledge production and sharing [CWANA; ESAP; Global; LAC; NAE; SSA].

Health and nutrition

The inter-linkages between health, nutrition, agriculture and AKST can constrain or facilitate reaching development and sustainability goals. Because multiple stressors affect these inter-linkages, a broad agroecosystem health approach is needed to identify appropriate AKST to increase food security and safety, decrease the incidence and prevalence of a range of infectious and chronic diseases, and decrease occupational exposures, injuries, and deaths.

Food security strategies require a combination of AKST approaches, including:

- Increasing the diversification of small-scale production and improve micronutrient intake;
- Increasing the efficiency and diversity of urban agriculture;
- Developing and deploying existing and new technologies for the production, processing, preservation, and distribution of food.

Food safety can be facilitated by effective, coordinated, and proactive national and international food safety systems, including:

- Enhancing public health and veterinary capacity, and legislative frameworks, for identification and control of biological and non-biological hazards;
- Vertical integration of the food chain to reduce the risks of contamination and alteration;
- Supporting the capacity of developing country governments, municipalities, and civil society organizations to develop systems for monitoring and controlling health risks along the entire food chain. One example is a battery of tests that municipalities could use to monitor pesticide residues on fruits and vegetables that are brought to market.

- Developing a system of global, national, and local AKST that can monitor developments and inform adequate and timely responses to the rapid evolution of pathogens.

The burden of emerging and reemerging diseases can be decreased by:
- Strengthening coordination between and the capacity of agricultural, veterinary, and public health systems;
- Integrating multi-sectoral policies and programs across the food chain to reduce the spread of infectious diseases;
- Developing and deploying new AKST to identify, monitor, control, and treat diseases; and
- Developing a system of global, national, and local AKST that can monitor developments and inform adequate and timely responses to the rapid evolution of pathogens and zoonotic outbreaks.

The burden of chronic diseases can be decreased by:
- Regulating food product formulation through legislation, international agreements and/or regulations for food labeling and health claims; and
- Creating incentives for the production and consumption of health-promoting foods.

Occupational health can be improved by:
- Developing and enforcing agriculture health and safety regulations;
- Enforcing cross-border issues such as illegal use of toxic agrichemicals; and
- Conducting health risk assessments that make explicit the trade-offs between maximizing benefits to livelihoods, the environment, and improving health.

Policies and institutional frameworks. Trends in the current burdens of the health risks associated with agriculture and AKST call for robust detection, surveillance, monitoring, and response systems to facilitate identification of the true burden of ill health and implementation of cost-effective, health-promoting strategies and measures. Persistent and substantial investment in capacity building are required to provide safe food of sufficient quantity, quality, and variety; reduce the burdens of obesity, other chronic diseases, and infectious diseases; and reduce agriculture-related environmental and occupational risks.

Equity
Science and technology (local and formal). Historically, formal AKST has privileged farmers with access to resources, services, capital and markets (e.g., men and non-indigenous groups), often creating greater inequalities in the rural sector. Additionally poor and marginalized groups have suffered disproportionately from environmental degradation [CWANA; LAC; SSA]. To acknowledge the distributional impact of AKST investments calls for conscious public policy choices to invest in AKST that addresses the needs of small-scale producers and improves equity [Global Chapters 3, 7]. This strategy recognizes that the short-term dollar rates of return may not as high as those of other investments

but that they can make a significant contribution to long-term poverty reduction.

For AKST to contribute to greater equity, investments are required for the development of appropriate technologies; access to education and research participation; new partnerships with a wider network of stakeholders; and models of learning, technology extension and facilitation for the poor and marginalized. Such investments are likely to improve access to sustainable technologies, credit and institutions (including property rights and tenure security) as well as to local, national, and regional markets for agricultural outputs [SR Part II: NRM].

Both formal and local AKST can add value to the full range of agricultural goods and services and help create economic instruments that promote an appropriate balance between private and public goods. At the farm, watershed, district and national scales, new methods may be needed to assess and improve the performance of farming systems in relation to the multiple functions of agriculture. Such efforts need to include a special emphasis on integrated water resource management for CWANA countries and other arid regions, and integrated soil management for SSA and other regions with highly degraded soils.

An environment in which formal science and technology and local and traditional knowledge are seen as part of an integral AKST system is most likely to increase equitable access to technologies to a broad range of producers [Global 3; SR Part II: NRM]. Options to improve this integration include moving away from a linear technology transfer approach that benefited relatively well-off producers of major cash crops but had little success for small-scale diversified farms and poor and marginalized groups and paid little attention to the multifunctionality of agriculture. Improvements are needed in engaging farmers in priority setting and funding decisions, and both in increasing collaboration with social scientists, and increasing participatory work in the core research institutions. Networks among small-scale producers contribute to the exchange of experience and AKST, as do inter- and multidisciplinary programs, cross-disciplinary learning and scientific validation, involving both research and non-research actors, and recognizing the cultural identity of indigenous communities.

Alternatives to traditional extension models include farmer field schools [SSA] and the *Campesino a Campesino* (Farmer to Farmer) Movement in LAC. However, such an integrated approach is unlikely to be embraced without complementary activities including developing in-country professional capacity for undertaking integrated approaches, methods for monitoring and evaluating these approaches, and ensuring a professional system that rewards participatory research in the top academic journals. A complementary option is to facilitate internal institutional learning and evaluation in AKST organizations, particularly as regards their impact on equity.

Policies and institutional frameworks. Key issues for improved performance include equitable access to and use of natural resources, systems of incentives and rewards for multifunctionality, including ecosystem services, and responding to the vulnerability of farming communities. Governance in AKST and related organizations are also important for the crucial role they play in democratization, decentralization and the integration of farmer concerns in the design of farmer services and agricultural industries. For example,

- AKST can assess IPR in terms of multifunctionality, consider issues of collective IPR and other non-IPR mechanisms such as prizes, cross-licensing and other means able to facilitate research and improve equity among regions. Legal frameworks can promote recognition of traditional knowledge associated with genetic resources and the equitable distribution of benefits derived there from among the custodians of these resources [Global Chapter 3]. Policies, including legal frameworks that regulate access to genetic resources and the equitable distribution of benefits generated by their use, can be implemented in ways that guarantee local communities access and the right to regulate the access of others. To date it is recognized that many poor regions bear the costs of protecting biodiversity and agricultural genetic diversity yet it is the global community who benefits from these practices. Thus, new national and international legal frameworks, in tandem with the development of institutions for benefit sharing, can ensure that local communities and individual countries control access to and benefit from local genetic resources as promoted in the Convention on Biological Diversity and as agreed in the International Treaty on Plant Genetic Resources for Food and Agriculture through its multilateral system of Access and Benefit Sharing.

- Large inequities in the tenure and access to land and water have exacerbated economic inequalities that still characterized many world regions in the world (e.g., LAC, SSA). Land reform, including improved tenure systems and equitable access to water are suggestive means to support sustainable management and simultaneously respond to social inequalities that inhibit economic development. Such initiatives are likely to reduce the displacement of small-scale farmers, *campesinos* and indigenous people to urban centers or to marginal lands in the agricultural frontier. Better understanding of the communal ownership, communal exchange and innovation mechanisms is needed. Overlapping formal and informal land rights that characterize some agricultural systems are central to strategies to reform land holdings and relations.

- In order to enhance a proper environment in which AKST contribute positively to development and sustainability goals, global equity can be enhanced by protecting small-scale farmers from unfair competition including from often subsidized commodities produced under conditions of economies of scale. Reasonable farm gate prices through equitable and fair access to markets and trade also are crucial for ensuring rural employment as well as improving livelihoods and food security. Such prices for small-scale holders can be achieved by eliminating commodity OECD agricultural subsidies to large industrialized farmers and dumping, and by not overexposing small-scale farmers to competition from industrial farmers before appropriate institutional frameworks and infrastructure are in place. They are also a condition for effective utilization of AKST. At the national and international level, governance mechanisms to respond to unfair competition and agribusiness accountability need to be implemented through, for example, anti-trust laws applied to financial institutions and the agrifood sector. One option might include creating or strengthening conditions that can guarantee farmers' rights to choose, select, and exchange seeds that are culturally and locally appropriate as well as to remove the monopoly from the privileges granted to breeders through Plant Breeders Rights through, for example, a compensatory liability regime.

- Global equity can be enhanced by improving small-scale farmers' access to international markets. The current trade environment in which agricultural subsidies and a history of public support to farming distort international prices for many key commodities can benefit from initiatives such as fair trade, organic certification, and sustainable timber certification. However, many schemes require additional skills that poorer farmers may have yet to access. In such circumstances, AKST can provide the training and support necessary to assist small-scale farmers in entering such markets.

- A direct connection between farmers and urban consumers (e.g., direct marketing and community-supported agriculture initiatives) can decrease the gap between the rural and urban sector and be of benefit to poor urban consumers. This can be accomplished by strengthening services, access to urban markets, centralized quality control, packaging and marketing to supply urban markets in the rural sector and particularly for small-scale producers. This approach is more likely to succeed if national farmers associations and their federations increase their role in national politics. AKST may also contribute to the development of urban and peri-urban agriculture focusing on the poorest urban sectors [LAC] as a means to enhance equity strengthen community organizations, support improved health, and promote food security as well as food sovereignty.

- When addressing issues of equity with respect to access to food, nutrition, health and a healthy environment, stakeholders can make use of established international treaties, agreements and covenants. For example the issue of hunger eradication can be supported by engaging the right to food as enshrined in Article 11 of the International Covenant on Economic, Social and Cultural Rights of the United Nations. This legal instrument, together with the International Covenants of Civil and Political Rights, is essential for putting into practice the principles set out in the Universal Declaration of Human Rights. In a culture of rights, states are obligated to take deliberate, concrete and non-discriminatory measures to eradicate hunger. To date, 146 countries are currently party to this covenant and 187 have signed the FAO Council's "voluntary guidelines for the progressive realization of the right to adequate food" [LAC].

- Despite their major and increasing contribution to agricultural production in several regions, particularly CWANA, LAC and SSA, women are marginalized with respect to access to education, extension services, and property rights, and are under-represented in agricultural science and technology teaching and development and extension services [Global Chapter 3]. Some women-oriented strategies, particularly increasing the functional literacy and general education levels of women, have already been proven to increase the likelihood of reaching development and sustainability goals [SSA and other regions]. Other actions, though not yet proven, include the reorientation of policies and programs to increase the participation and physical presence of women in leadership, decision-making, and implementation positions. Specific actions to mainstream women's involvement include encouraging women by generating stimuli and opportunities to study agricultural sciences and economics, and also to ensure that activities such as extension, data collection, and enumeration involve women as providers as well as recipients. Farmer research groups, too, have proven more successful in reaching women farmers than traditional extension activities [SSA] suggesting that similar approaches may be needed to incorporate marginalized groups—the landless, pastoralists, and seasonal and longer-term migrants—into education and policy making institutions.

- Participation in and democratization of AKST processes helps to integrate sectors (i.e., developing networks), which have been excluded [Global Chapter 3]. These processes include improved access to information and institutional support to and the development of education and training in ways that incorporate the participation of civil society as ones means to guarantee transparency and accountability. A key point is helping youth to become involved in agriculture and of making it an attractive work activity compared with urban possibilities. Long-term investment in farmer education, especially for women and youth, the empowerment of farmers as vocal partners in business and IPR development and other legal framework, and strengthening civil society organizations.

- Improving equity requires synergy among various development actors, including farmers, agricultural workers, banks, civil society organizations, commercial companies, and public agencies [Global Chapter 3]. Stakeholder involvement is also crucial in decisions about infrastructure, tariffs, and the internalization of social and environmental costs. Women and other historically marginalized actors (local/indigenous community members, farm workers, etc.) need to have an active role in problem identification (determining research questions, extension objectives, etc.) and policy and project design. New modes of governance to develop innovative local networks and decentralized government, focusing on small-scale producers and the urban poor (urban agriculture) will help to create and strengthen synergetic and complementary capacities [LAC].

Investments

The contribution of AKST to the achievement of development and sustainability goals would entail increased funds and more diverse funding mechanisms for agricultural research and development and associated knowledge systems. These could include:

- Public investments to serve global, regional and local public goods, addressing strategic issues such as food security and safety, climate change and sustainability that do not attract private funding. More efficient use of increasingly scarce land, water and biological resources would need public investment in legal and management capabilities.
- Public investment to support effective change in agricultural knowledge systems directed to:
    - promote interactive knowledge networks (associating farmers, farmers communities, scientists, industrial and actors in other knowledge areas) and improve access for all actors to information and communication technologies;
    - support ecological, evolutionary, food, nutrition, social and complex systems' sciences and the promotion of effective interdisciplinarity;
    - establish capacities and facilities to offer life-long learning opportunities to those involved in the agrifood arena.
- Public-private partnerships for improved commercialization of applied knowledge and technologies and joint funding of AKST, where market risks are high and where options for widespread utilization of knowledge exists;
- Adequate incentives and rewards to encourage private and civil society investments in AKST contributing to development and sustainability goals.

There are many options to target investments to contribute to the development and sustainability goals. Options have to be examined with high consideration of local and regional, social, political and environmental contexts, addressing goals such as:

- *Poverty, livelihoods and food security.* AKST investments can increase the sustainable productivity of major subsistence foods including orphan crops that are grown and/or consumed by the poor. Investments could also be targeted for institutional change and policies that can improve access of poor people to food, land, water, seeds, germplasm and improved technologies, particularly in value chain addition technologies such as quality processing of agricultural products
- *Environmental sustainability.* Increased investments are needed in AKST that can improve the sustainability of agricultural systems and reduce their negative environmental effects with particular attention to alternative production systems, e.g., organic and low-input systems; reduce greenhouse gas emissions from agricultural practices; reduce the vulnerability of agroecological systems to the projected changes in climate and climate variability (e.g., breeding for temperature and pest tolerance); understanding the relationship between ecosystem services provided by agricultural systems and their relationships to human well-being; economic and non-economic valuation of ecosystem services; improving water use efficiency and reducing water pollution; developing biocontrols of current and emerging pests and pathogens, and biological substitutes for agrochemicals; and reducing the dependency of the agricultural sector on fossil fuels.

- *Human health and nutrition.* Major public and private AKST investments will be needed to contribute to: the reduction of chronic diseases through scientific programs and legislation related to healthy diets and food product formulations; the improvement of food safety regulations in an increasingly commercialized and globalized food industry; the control and management of infectious diseases, through the development of new vaccines, global surveillance, monitoring and response systems and effective legal frameworks. In addition, investments are needed in science and legislation covering occupational health issues such as pesticide use and safety regulations (including child labor laws).

- *Equity.* Preferential investments in equitable development, as in literacy, education and training, that contribute to reducing ethnic, gender, and other inequities would advance the development and sustainability goals. Measurements of returns to investments require indices that give more information than GDP, and that are sensitive to environmental and equity gains. The use of inequality indices for screening AKST investments and monitoring outcomes strengthens accountability. The Gini-coefficient could, for example, become a public criterion for policy assessment, in addition to the more conventional measures of growth, inflation and environment.

In many developing countries, it may be necessary to complement these investments with increased and more targeted investments in rural infrastructure, education and health and to strengthen capacity in core agricultural and related sciences.

In the face of new global challenges, there is a urgent need to strengthen, restructure and possibly establish new intergovernmental, independent science-based networks to address such issues as climate forecasting for agricultural production; human health risks from emerging diseases such as avian flu; reorganization of livelihoods in response to changes in agricultural systems (population movements); food security; and global forestry resources.

Part II: Themes

Bioenergy

Writing team: Patrick Avato (Germany/Italy), Rodney J. Brown (USA), Moses Kairo (Kenya)

Bioenergy has recently received considerable public attention. Rising costs of fossil fuels, concerns about energy security, increased awareness of climate change, domestic agricultural interests and potentially positive effects for economic development all contribute to its appeal for policy makers and private investors. Bioenergy as defined in the IAASTD covers all forms of energy derived from biomass, e.g., plants and plant-derived materials. Bioenergy is categorized as traditional or modern, depending on the history of use and technological complexity. Traditional bioenergy includes low technology uses including direct combustion of firewood, charcoal or animal manure for heat generation. Modern bioenergy is comprised of electricity, light and heat produced from solid, liquid or gasified biomass and liquid biofuels for transport. Liquid biofuels for transport can be categorized as first generation, produced from starch, sugar or oil containing agricultural crops, or next generation. Next generation (also referred to as second, third or fourth generation) biofuels are produced from a variety of biomass materials, e.g., specially grown energy crops, agricultural and forestry residues and other cellulosic material [CWANA Chapter 2; Global Chapters 3, 6; NAE Chapter 4].

As biomass feedstocks are widely available, bioenergy offers an attractive complement to fossil fuels and thus has potential to alleviate concerns of a geopolitical and energy security nature. However, only a small part of globally available biomass can be exploited in an economically, environmentally and socially sustainable way. Currently, about 2.3% of global primary energy is supplied by modern sources of bioenergy such as ethanol, biodiesel, or electricity and industrial process heat [Global Chapter 3].

The economics of bioenergy, and particularly the positive or negative social and environmental externalities, vary strongly, depending on the source of biomass, type of conversion technology and on local circumstances and institutions. Many questions in development of bioenergy will require further research. Agricultural knowledge, science, and technology (AKST) can play a critical role in improving benefits and reducing potential risks and costs but complementary efforts are needed in the areas of policies, capacity building, and investment to facilitate a socially, economically, and environmentally sustainable food, feed, fiber, and fuels economy. Specific options and challenges associated with the different categories are discussed in the following

section. Aspects that are crosscutting are discussed in a separate section.

Traditional Bioenergy
Millions of people in developing countries depend on traditional biofuels for their most basic cooking and heating needs (e.g., wood fuels in traditional cook stoves or charcoal). Dependence on traditional bioenergy is highly correlated with low income levels and is most prevalent in sub-Saharan Africa and South Asia due to a lack of affordable alternatives. In some countries, the share of biomass in energy consumption can reach up to 90%. Within countries, the use of biomass is heavily skewed toward the lowest income groups and rural areas [CWANA Chapter 2; Global Chapter 3; SSA Chapter 2].

Reliance on traditional bioenergy can stifle development by posing considerable environmental, health, economic and social challenges. Traditional biomass is usually associated with time consuming and unsustainable harvesting, hazardous pollution and low end-use efficiency, and in the case of manure and agricultural residues depletion of soil by removal of organic matter and nutrients. Collecting fuel is time-consuming, reducing the time that can be devoted to productive uses including farming and education. Air pollution from biomass combustion leads to asthma and other respiratory problems which lead to 1.5 million premature deaths per year[7] [Global Chapter 3; SSA Chapter 2]. Efforts in the past at making available improved and more efficient traditional bioenergy technologies (e.g., improved cook stoves) have led to mixed results. New and improved efforts and approaches are therefore needed that build on and expand these efforts. Moreover, other options must be explored to expand the availability and use of modern energy solutions. Such technologies differ widely from each other in terms of economic, social and environmental implications and may include fossil fuels, extensions of electricity grids, and forms of distributed energy including modern forms of bioenergy (see section on bioelectricity and bioheat).

First Generation Biofuels
First generation biofuels consist today predominantly of bioethanol and biodiesel, even though other fuels such as methanol, propanol and butanol may play a larger role in the future. Produced from agricultural crops such as maize

---

[8] This number includes deaths caused by the combustion of coal in the homestead.

From biomass to energy consumption

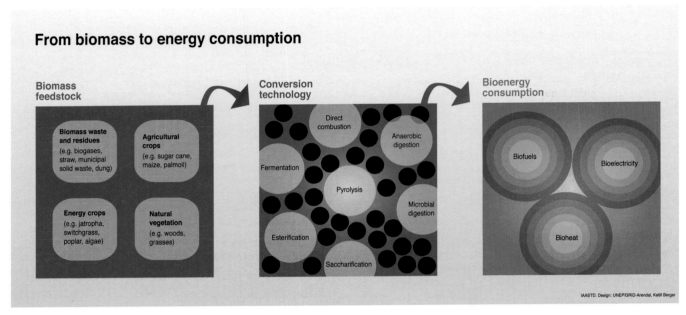

Figure SR-BE1. *From biomass to energy consumption.*

and other grains, sugar cane, soybeans, cassava, rapeseed, and oil palm, production of bioethanol and biodiesel has been growing fast in recent years, albeit from a low base—together they contributed about 1% of global transport fuels in 2005. Fast growth rates are mainly due to biofuel support policies that have been developed in many countries around the world in the hope of furthering rural job creation and economic development, mitigating climate change and improving energy security [ESAP Chapter 4; NAE Chapter 2; SSA Chapter 2].

The most important factors determining economic competitiveness of first generation biofuels are (1) price of feedstock, (2) value of byproducts, (3) conversion technology, and (4) price of competing fuels. Each of these variables varies over time and place. Currently first generation biofuels are economically competitive with fossil fuels only in the most efficient feedstock producer markets during times of favorable market conditions, e.g., in Brazil when feedstock prices are low and fossil fuel prices high. Consistently high oil prices at levels seen in the recent past would improve economic competitiveness also in other regions. The economics of liquid biofuels may be more favorable in remote regions where energy access and agricultural exports are complicated by high transport costs. Land-locked developing countries, islands, and remote regions within countries may fall into this category if they can make available sufficient and cheap feedstock without threatening food security [Global Chapters 3, 6; NAE Chapter 4].

In addition to these economic factors, the value of 1st generation biofuels is also affected by energy security concerns and environmental and social benefits and costs. From an environmental perspective, there is considerable debate over whether first generation biofuels, especially bioethanol, yield more energy than is needed for their production and their level of greenhouse gas emissions. Both issues are related and the debate is caused by differences in life cycle emissions measurement methodologies and the strong effect of specific local circumstances, such as type of feedstock, original use of agricultural land, mechanization of production and fertilizer use. Generally, assuming feedstocks are produced on agricultural land and do not induce deforestation, crops produced with few external inputs (fertilizers, pesticides, etc.), such as rain fed sugarcane in Brazil, perform significantly better than high-input crops such as maize in North America. Consequently, whether biofuels are a viable option for climate change mitigation depends on the emissions reductions that can realistically be achieved as well as relative costs compared to other mitigation alternatives. Apart from GHG considerations, considerable environmental costs may be associated with large increases in biofuels production. For example, it is feared that the increased demand for limited agricultural production factors (e.g., land and water) will lead to a conversion of pristine biodiverse ecosystems to agricultural land (e.g., deforestation) and depletion of water resources—instances of this happening are already apparent in different regions, e.g., draining of peat land in Indonesia and clearing of the Cerrado in Brazil [Global Chapters 4, 6; NAE Chapter 4].

The related social and economic effects are complex. Increased demand can lead to higher incomes for those engaged in feedstock production and ancillary industries such as biofuels conversion or processing of biofuel by-products (e.g., cakes), potentially contributing to economic development. Conversely, competition for limited land and water resources inevitably leads to higher food prices hurting buyers of food, including food processing and livestock industries and—very importantly with regard to hunger and social sustainability—poor people. Moreover, small-scale farmers may be marginalized or pushed off their lands if they are not protected and brought into production schemes. In the medium to long term the effects on food prices may decrease as economies react to higher prices (adapting production patterns and inducing investments) and technologies improve. Consequently, the social and economic effects have strong distributional impacts within societies, between different stakeholders and over time. Institutional arrangements strongly influence the distribution of these effects, e.g., between small and large producers and between men and women [Global Chapter 6].

In addition to the direct effects of biofuel production, policies employed to promote them create their own costs and benefits. As first generation biofuels have rarely been economically competitive with petroleum fuels, production in practically all countries is promoted through a complex set of subsidies and regulations. In addition to the direct budgetary costs of such subsidies, policies in most countries contain market distortions such as blending mandates, trade restrictions and tariffs that create costs through inefficiencies. This undermines an efficient allocation of biofuel production in the countries with the largest potential and cheapest costs and creates costs for consumers.

Liberalizing biofuel trade through the reduction of trade restrictions and changes in the trade classification of ethanol and biodiesel would promote a more efficient allocation of production in those countries that have a comparative advantage in feedstock production and fuel conversion, respectively. However, it is not clear how resource-poor small-scale farmers could benefit from this. Moreover, unless environmental and social sustainability is somehow ensured, negative effects such as deforestation, unsustainable use of marginal lands and marginalization of small-scale farmers risk being magnified. Sustainability standards and voluntary approaches are the most frequently discussed options for ensuring socially and environmentally sustainable biofuel production. However, there is currently no international consensus on what such schemes should encompass, whether they could effectively improve sustainability or even whether they should be developed at all [Global Chapter 7].

AKST can play a role in improving the balance of social, environmental and economic costs and benefits, albeit within limits. R&D on increasing biofuel yields per hectare while reducing agricultural input requirements by optimizing cropping methods, breeding higher yielding crops and employing local plant varieties offers considerable potential. Both conventional breeding and genetic engineering are being employed to further enhance crop characteristics such as starch, sugar, cellulose or oil content to increase fuel-producing capacity [Global Chapter 6]. A variety of crops and cropping methods in different countries are believed to hold large yield potential, each adapted to specific environments, but more research is needed to develop this potential.

Next Generation Biofuels
The development of new biofuel conversion technologies, so-called next generation biofuels, has significant potential. Cellulosic ethanol and biomass-to-liquids (BTL) technologies, the two most prominent technologies, allow the conversion into biofuels not only of the glucose and oils retrievable today but also of cellulose, hemi-cellulose and even lignin—the main building blocks of most biomass. Thereby, more abundant and potentially cheaper feedstocks such as residues, stems and leaves of crops, straw, urban

wastes, weeds and fast growing trees could be converted into biofuels. Further in the future is the possibility of using sources, such as algae or cyanobacteria intensively cultivated in ponds or bioreactors in saline water using industrial carbon dioxide. Research is also focusing on integrating the production of next generation biofuels with the production of chemicals, materials and electricity. These so-called biorefineries could improve production efficiency, GHG balances and process economics.

On the one hand, the wide variety of potential feedstocks and high conversion efficiencies of next generation biofuels could dramatically reduce land requirements per unit of energy produced, thus mitigating the food price and environmental pressures of first generation biofuels. Moreover, lifecycle greenhouse gas emissions could be reduced relative to first generation biofuels. On the other hand, there are concerns about unsustainable harvesting of agricultural and forestry residues and the use of genetically engineered crops and enzymes. However, as next generation biofuels are still nascent technologies, these economic, social and environmental costs and benefits are still very uncertain [Global Chapters 6, 7; NAE Chapter 4].

Several critical steps have to be overcome before next generation biofuels can become an economically viable source of transport fuels. It is not yet clear when these breakthroughs will occur and what degree of cost reductions they will be able to achieve in practice. Moreover, while some countries like South Africa, Brazil, China and India may have the capacity to actively engage in advanced domestic biofuels R&D efforts, high capital costs, large economies of scale, a high degree of technical sophistication and IPR issues make the production of next generation biofuels problematic in the majority of developing countries, even if the technological and economic hurdles can be overcome in industrialized countries. Arrangements are therefore needed to address these issues in developing countries and for small farmers [Global Chapters 6, 8].

Bioelectricity and Bioheat
Bioelectricity and bioheat are produced mostly from biomass wastes and residues. Use of both small-scale biomass digesters and larger-scale industrial applications has expanded in recent decades. Generation of electricity (44 GW-24 GW in developing countries—in 2005 or 1% of total electricity consumption) and heat (220 GWth in 2004) from biomass is the largest non-hydro source of renewable energy, mainly produced from woods, residues and wastes.

The major biomass conversion technologies are thermochemical and biological. The thermo-chemical technologies include direct combustion of biomass (either alone or co-fired with fossil fuels) and gasification (to producer gas). The biological technologies include the anaerobic digestion of biomass to yield biogas (a mixture primarily of methane and carbon dioxide). Household-scale biomass digesters that operate with local organic wastes like animal manure can generate energy for cooking, heating and lighting in rural homes and are widespread in China, India and Nepal, with the organic sludge and effluents returned to the fields. However their operation can sometimes pose technical, maintenance and resource challenges (e.g., water requirements of digesters). Industrial-scale units are less prone to technical problems and are increasingly widespread in some developing countries, especially in China. Similar technologies are also employed in industrialized countries, mostly to capture environmentally problematic methane emissions (e.g., at landfills and livestock holdings) and produce energy.

Some forms of bioelectricity and bioheat can be economically competitive with other off-grid energy options such as diesel generators, even without taking into consideration potential non-market benefits such as GHG emissions reductions, and therefore are viable options for expanding energy access in certain settings. The largest potential lies with the production of bioelectricity and heat when technically mature and reliable generators have access to secure supply of cheap feedstocks and capital costs can be spread out over high average electricity demand. This is sometimes the case on site or near industries that produce biomass wastes and residues and have their own steady demand for electricity, e.g., sugar, rice and paper mills. Environmentally and socially, bioelectricity and heat are most often less problematic than liquid biofuels for transport because they are predominantly produced from wastes, residues and sustainable forestry. In these cases significant GHG emission reductions can be achieved, even when biomass is co-fired with coal, and food prices are unlikely to be affected. The economics as well as environmental effects are particularly favorable when operated in combined heat and electricity mode, which is increasingly being employed in various countries, e.g., during harvesting season Mauritius meets 70% of electricity needs from sugarcane bagasse cogeneration. However, particulate emissions from smoke stacks are of considerable concern. Biomass digesters and gasifiers are more prone to technical failures than direct combustion facilities, especially when operated in small-scale applications without proper maintenance and experiences with their application vary considerably [ESAP Chapter 4; Global Chapters 3, 5, 6; SSA Chapter 2].

Small-scale applications for local use of first generation biofuels can sometimes offer interesting alternatives for electricity generation that do not necessarily produce the negative effects of large-scale production due to more contained demands on land, water and other resources. Biodiesel has special potential in small-scale applications, as it is less technology and capital intensive to produce than ethanol, although methanol requirements for its production can pose a challenge. Unrefined bio-oils for stationary uses are even less technology intensive to produce and do not require methanol. However, engines for power generation and water pumping have to be adapted for their use. Local stationary biofuel schemes may offer particular potential for local communities when they are integrated in high intensity small-scale farming systems that allow an integrated production of food and energy crops. These options are being analyzed in several countries, e.g., focusing on Jatropha and Pongamia as a feedstock, but evidence on their potential is not yet conclusive [CWANA Chapter 2; Global Chapter 6; NAE Chapter 5].

Several actions can be undertaken to promote a better exploitation of bioelectricity and bioheat potential [Global Chapter 7].

- *Promoting R&D:* Improving operational stability and reducing capital costs promises to improve the attractiveness of bioenergy, especially of small and medium-scale biogas digesters, thermo-chemical gasifiers and stationary uses of unrefined vegetable oils. More research is also needed on assessing the costs and benefits to society of these options, taking into consideration also other energy alternatives [Global Chapter 6].

- *Development of product standards and dissemination of knowledge:* A long history of policy failures and a wide variety of locally produced generators with large differences in performance have led to considerable skepticism about bioenergy in many countries. The development of product standards, as well as demonstration projects and better knowledge dissemination, can contribute to increase market transparency and improve consumer confidence.

- *Local capacity building:* Experience of various bioenergy promotion programs has shown that proper operation and maintenance are key to success and sustainability of low-cost and small-scale applications. Therefore, local consumers and producers need to be closely engaged in the development as well as the monitoring and maintenance of facilities.

- *Access to finance:* Compared to other off-grid energy solutions, bioenergy often exhibits higher initial capital costs but lower long-term feedstock costs. This cost structure often forces poor households and communities to forego investments in modern bioenergy—even in cases when levelized costs are competitive and payback periods short. Improved access to finance can help to reduce these problems.

Cross-cutting Issues

Food prices. The diversion of agricultural crops to fuel can negatively affect hunger alleviation throughout the world in the short to medium term, even though price increases may be mitigated in the long term. This risk is particularly high for first generation biofuels for transport due to their very large demands for agricultural crops. Price increases can be caused directly, through the increase in demand for feedstocks, or indirectly, through the increase in demand for the factors of production (e.g., land, water), so the use of non-food crops is unlikely to alleviate these concerns. More research is needed to assess these risks and their effects but it is evident that poor net buyers of food and food-importing developing countries are particularly affected.

Environment. The large demands for additional agricultural and forestry products for bioenergy can also cause important environmental effects. Again, because of the large additional demands for agricultural feedstocks, first generation biofuels create the largest potential problems including pushing more ecologically fragile and valuable lands into production and depleting and contaminating water resources. Moreover, some of the fast growing crops promoted for bioenergy production raise environmental (e.g., their resemblance with weeds) and social concerns. On the other hand, bioenergy can positively contribute to climate change mitigation, although this potential differs strongly from case to case and costs have to be compared to other mitigation options.

Institutional arrangements. Institutional arrangements and power relationships strongly impact the ability of different stakeholders to participate in bioenergy production and consumption and the distribution of costs and benefits. The current weaknesses in institutional links and responsibilities between the various sectors involved in the policy and

technology of agriculture as an energy consumer and producer will have to be overcome through local, national and regional frameworks.

Integrated analysis. The economics of bioenergy as well as positive and negative environmental and social effects are highly complex, depend considerably on particular circumstances and have important distributional implications. Consequently, decision makers need to carefully weigh full social, environmental and economic costs of the targeted form of bioenergy and of the envisaged support policy against realistically achievable benefits and other energy alternatives.

Biotechnology

Writing Team: Jack Heinemann (New Zealand), Tsedeke Abate (Ethiopia), Angelika Hilbeck (Switzerland), Doug Murray (USA)

Biotechnology[8] is defined as "any technological application that uses biological systems, living organisms, or derivatives thereof, to make or modify products or processes for a specific use." In this inclusive sense, biotechnology can include anything from fermentation technologies (e.g., for beer making) to gene splicing. It includes traditional and local knowledge (TLK) and the contributions to cropping practices, selection and breeding of plants and animals made by individuals and societies for millennia [CWANA Chapter 1; Global Chapter 6]. It would also include the application of tissue culture and genomic techniques [Global Chapter 6] and marker assisted breeding or selection (MAB or MAS) [Global Chapter 5, 6; NAE Chapter 2] to augment natural breeding.[9]

Modern biotechnology is a term adopted by international convention to refer to biotechnological techniques for the manipulation of genetic material and the fusion of cells beyond normal breeding barriers[9] [Global Chapter 6]. The most obvious example is genetic engineering to create genetically modified/engineered organisms (GMOs/GEOs) through "transgenic technology" involving the insertion or deletion of genes. The word "modern" does not mean that these techniques are replacing other, or less sophisticated, biotechnologies.

Conventional biotechnologies, such as breeding techniques, tissue culture, cultivation practices and fermentation are readily accepted and used. Between 1950 and 1980, prior to the development GMOs, modern varieties of wheat may have increased yields up to 33% even in the absence of fertilizer. Even modern biotechnologies used in containment have been widely adopted. For example, the industrial enzyme market reached US$1.5 billion in 2000.

Biotechnologies in general have made profound contributions that continue to be relevant to both big and small farmers and are fundamental to capturing any advances derived from modern biotechnologies and related nanotechnologies[10] [Global Chapter 3, 5, 6]. For example, plant breeding is fundamental to developing locally adapted plants whether or not they are GMOs. These biotechnolo-gies continue to be widely practiced by farmers because they were developed at the local level of understanding and are supported by local research.

Much more controversial is the application of modern biotechnology outside containment, such as the use of GM crops. The controversy over modern biotechnology outside of containment includes technical, social, legal, cultural and economic arguments. The three most discussed issues on biotechnology in the IAASTD conceredt:

- Lingering doubts about the adequacy of efficacy and safety testing, or regulatory frameworks for testing GMOs [e.g., CWANA Chapter 5; ESAP Chapter 5; Global Chapter 3, 6; SSA 3];
- Suitability of GMOs for addressing the needs of most farmers while not harming others, at least within some existing IPR and liability frameworks [e.g., Global Chapter 3, 6];
- Ability of modern biotechnology to make significant contributions to the resilience of small and subsistence agricultural systems [e.g., Global Chapter 2, 6].

Some controversy may in part be due to the relatively short time modern biotechnology, particularly GMOs, has existed compared to biotechnology in general. While many regions are actively experimenting with GMOs at a small scale [e.g., ESAP Chapter 5; SSA Chapter 3], the highly concentrated cultivation of GM crops in a few countries (nearly three-fourths in only the US and Argentina, with 90% in the four countries including Brazil and Canada) is also interpreted as an indication of a modest uptake rate [Global Chapter 5, 6]. GM crop cultivation may have increased by double digit rates for the past 10 years, but over 93% of cultivated land still supports conventional cropping.

The pool of evidence of the sustainability and productivity of GMOs in different settings is relatively anecdotal, and the findings from different contexts are variable [Global Chapter 3, 6], allowing proponents and critics to hold entrenched positions about their present and potential value. Some regions report increases in some crops [ESAP Chapter 5] and positive financial returns have been reported for GM cotton in studies including South Africa, Argentina, China, India and Mexico [Global Chapter 3; SSA Chapter 3]. In contrast, the US and Argentina may have slight yield declines in soybeans, and also for maize in the US [references in Global Chapter 3]. Studies on GMOs have also shown the potential for decreased insecticide use, while others show increasing herbicide use. It is unclear whether detected benefits will extend to most agroecosystems or be sustained

---

[9] See definition in Executive Summary.

[10] These are provided as examples and not comprehensive descriptions of all types of modern biotechnology (see Fig. SR-BT1).

[11] Specifically those nanotechnologies that involve the use of living organisms or parts derived thereof.

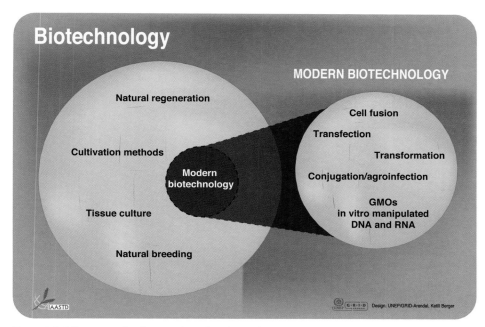

Figure SR-BT1. *Biotechnology and modern biotechnology defined.*

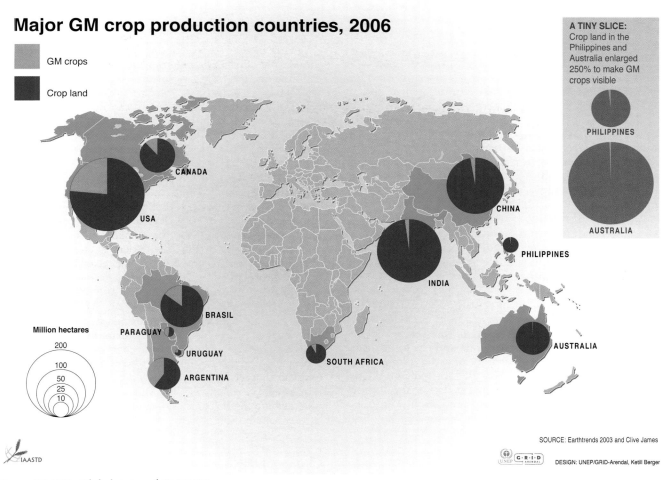

Figure SR-BT2. *Global status of GM 2006.*

Land area: Conventional and GM crops

GM share of total (per cent)

■ Conventional ■ GM

Growth in GM agriculture

Global GM plantings by country ('000 hectares)

SOURCE: Clive James and Wenzel, G ("2006) Appl. Microbiol. Biotechnol. volume 70, p. 642–650

IAASTD. Design: UNEP/GRID-Arendal, Ketill Berger

Figure SR-BT3. *Agricultural land (1996-2000) by GM and conventional crop plantings: keeping scale in perspective.*

in the long term as resistances develop to herbicides and insecticides [Global Chapter 3].

IPR frameworks need to evolve to increase access to proprietary biotechnologies, especially modern biotechnology, and address new liability issues for different sectors of producers. The use of IPR to increase investment in agriculture has had an uneven success when measured by type of technology and country. In developing countries especially, too often instruments such as patents are creating prohibitive costs, threatening to restrict experimentation by the individual farmer or public researcher while also potentially undermining local practices that enhance food security and economic sustainability. In this regard, there is particular concern about present IPR instruments eventually inhibiting seed-savings and exchanges.

Modern biotechnology has developed in too narrow a context to meet its potential to contribute to the small and subsistence farmer in particular [NAE Chapter 6, SDM]. As tools, the technologies in and of themselves cannot achieve sustainability and development goals [CWANA Chapter 1; Global Chapter 2, 3]. For example, a new breeding technique or a new cultivar of rice is not sufficient to meet the requirements of those most in need; the grain still has to be distributed. Dissemination of the technique or variety alone would not reduce poverty; it must be adapted to local conditions. Therefore, it is critical for policy makers to holistically consider biotechnology impacts beyond productivity and yield goals, and address wider societal issues of capacity building, social equity and local infrastructure [SSA Chapter 3].

Challenge: Biotechnology for Development and Sustainability Goals

Biotechnology in general, and modern biotechnology in particular, creates both costs and benefits [CWANA Chapter 5;

ESAP Chapter 5; Global Chapter 3], depending on how it is incorporated into societies and ecosystems and whether there is the will to fairly share benefits as well as costs. For example, the use of modern plant varieties has raised grain yields in most parts of the world, but sometimes at the expense of reducing biodiversity or access to traditional foods [Global Chapter 3]. Neither costs nor benefits are currently perceived to be equally shared, with the poor tending to receive more of the costs than the benefits [Global Chapter 2].

Hunger, nutrition and health

Biotechnologies affect human health in a variety of ways. The use of DNA-based technologies, such as microchips, for disease outbreak surveillance and diagnostics can realistically contribute to both predicting and curtailing the impacts of infectious diseases [NAE Chapter 6]. The application of these technologies would serve human health objectives both directly and indirectly, because they could be applied to known human diseases and to plant and animal diseases that might be the source of new human diseases or which could reduce the quantity or quality of food.

Other products of modern biotechnology, for example GMOs made from plants that are part of the human food supply but developed for animal feed or to produce pharmaceuticals that would be unsafe as food, might threaten human health [Global Chapters 3, 6]. Moreover, the larger the scale of bio/nanotechnology or product distribution, the more challenging containment of harm can become [Global Chapter 6].

All biotechnologies must be better managed to cope with a range of ongoing and emerging problems [SSA Chapter 3]. Holistic solutions may be slowed, however, if GMOs are seen as sufficient for achieving development and sustainability goals and consequently consume a disproportionate level of funding and attention. To use GMOs or not is a decision that requires a comprehensive understanding of the products, the problems to be solved and the societies in which they may be used [CWANA Chapter 5]. Thus, whatever choices are made, the integration of biotechnology must be within an enabling environment supported by local research [Global Chapter 6] and education that empowers local communities [CWANA Chapter 1].

Social equity

Two framing perspectives on how best to put modern biotechnology to work for achieving sustainability and development goals are contrasted in the IAASTD. The first perspective [e.g., see Global Chapter 5] argues that modern biotechnology is overregulated and this limits the pace and full extent of its benefits. According to the argument, regulation of biotechnology may slow down the distribution of products to the poor [Global Chapter 5].

The second perspective says that the largely private control of modern biotechnology [Global Chapter 5] is creating both perverse incentive systems, and is also eroding the public capacity to generate and adopt AKST that serves the public good [e.g., see Global Chapters 2, 7]. The integration of biotechnology through the development of incentives for private (or public-private partnership) profit has not been successfully applied to achieving sustainability and development goals in developing countries [Global Chapter 7], especially when they include the success of emerging and small players in the market. Consolidation of larger economic units [CWANA Chapter 1; Global Chapter 3; NAE Chapters 2, 6] can limit agrobiodiversity [Global Chapter 3] and may set too narrow an agenda for research [Global Chapters 2, 5]. This trend might be slowed through broadening opportunities for research responsive to local needs.

The rise of IPR frameworks since the 1970s, and especially the use of patents since 1980, has transformed research in and access to many products of biotechnology [Global Chapter 2; NAE Chapter 2]. Concerns exist that IPR instruments, particularly those that decrease farmers' privilege, may create new hurdles for local research and development of products [Global Chapters 2, 6; SSA Chapter 3]. It is unlikely, therefore, that over regulation per se inhibits the distribution of products from modern biotechnology because even if safety regulations were removed, IPR would still likely be a significant barrier to access and rapid adoption of new products. This may also apply to the future development of new GM crops among the largest seed companies, with costs incurred to comply with IP requirements already exceeding the costs of research in some cases [Global Chapters 6, 7].

Products of biotechnology, both modern and conventional, are frequently amenable to being described as IP and increasingly being sold as such, with the primary holders of this IP being large corporations that are among those most capable of globally distributing their products [Global Chapter 2]. Even under initiatives to develop "open source" biotechnology or return some IP to the commons, the developers may have to adequately document the IP to prevent others from claiming it and restricting its use in the future.

This ability to develop biotechnologies to meet the needs of IP protection goals may undervalue the past and present contribution by farmers and societies to the platform upon which modern biotechnology is built [ESAP Chapter 5; Global Chapters 2, 6, 7]. It is not just the large transnational corporations who are interested in retaining control of IP. Public institutions, including universities, are becoming significant players and in time, holders of TLK may also [Global Chapter 7].

IP protected by patents can be licensed for use by others. Currently it is contracts and licenses [Global Chapter 2] that dominate the relationship between seed developers and farmers [Global Chpater 2]. For example, farmers and CGIARs enter into contracts and material transfer agreements (MTAs) with a seed company, or a community-based owner of TK. These contracts can help resolve some access issues, but can simultaneously create other legal and financial problems that transcend easy fixes of patent frameworks alone [Global Chapters 2, 5].

Technical and Intensification Issues

Since agriculture (excluding wild fisheries) already uses nearly 40% of the Earth's land surface [Global Chapter 7], biotechnology could contribute to sustainability and development goals if it were to help farmers of all kinds produce more from the land and sea already in use, rather than by producing more by expanding agricultural land [SSA Chapter 1]. In addition to meeting future food needs, agriculture is increasingly being considered as an option to meet energy needs [Global Chapter 6], which exacerbates the pressures on yield [ESAP Chapter 5]. Food security, however, is a multi-dimensional challenge, so the demands on biotechnology in the long term will extend far beyond just increasing yield [NAE Chapter 6, SDM].

Agroecosystems

How agriculture is conducted influences what and how much a society can produce. Biotechnology and the production system are inseparable, and biotechnology must work with the best production system for the local community [ESAP Chapter 5]. For example, agroecosystems of even the poorest societies have the potential through ecological agriculture and IPM to meet or significantly exceed yields produced by conventional methods, reduce the demand for land conversion for agriculture, restore ecosystem services (particularly water), reduce the use of and need for synthetic fertilizers derived from fossil fuels, and the use of harsh insecticides and herbicides [Global Chapters 3, 6, 7]. Likewise, how livestock are farmed must also suit local conditions [CWANA Chapter 1]. For example, traditional "pastoral societies are driven by complex interactions and feedbacks that involve a mix of values that includes biological, social, cultural, religious, ritual and conflict issues. The notion that sustainability varies between modern and traditional societies needs to be" generally recognized [Global Chapter 6]. It may not be enough to use biotechnology to increase the number or types of cattle, for instance, if this reduces local genetic diversity or ownership, the ability to secure the best adapted animals, or they further degrade ecosystem services [CWANA Chapters 1, 5; Global Chapter 7].

Agroecosystems are also vulnerable to events and choices made in different systems. Some farming certification systems, e.g., organic agriculture, can be put at risk by GMOs, because a failure to segregate them can undermine market certifications and reduce farmer profits [Global Chapter 6]. Seed supplies and centers of origin may be put at risk when they become mixed with unapproved or regulated articles in source countries [Global Chapter 3].

Trees and crops

Plant breeding and other biotechnologies (excluding transgenics discussed below) have made substantial historical contributions to yield [Global Chapter 3]. While yield may have "topped out" under ideal conditions [Global Chapter 3], in developing countries the limiting factor has been access to modern varieties and inputs instead of an exhaustion of crop trait diversity [Global Chapter 3], and therefore plant breeding remains a fundamental biotechnology for contributing to sustainability and development goals.

Biotic and abiotic stresses, e.g., plant pathogens, drought and salinity, pose significant challenges to yield. These challenges are expected to increase with the effects of urbanization, the conversion of more marginal lands to agricultural use [SSA Chapter 1], and climate change [CWANA Chapter 1; Global Chapter 7; SSA Chapter 1]. Adapting new cultivars to these conditions is difficult and slow, but it is again plant breeding perhaps complemented with MAS, that is expected to make the most substantial contribution [Global Chapters 3, 6]. Genetic engineering also could be used to introduce these traits [Global Chapter 5; NAE Chapter 6]. It may be a way to broaden the nutritional value of some crops [ESAP Chapter 5]. If GM crops were to increase productivity and prevent the conversion of land to agricultural use, they could have a significant impact on conservation [Global Chapter 5]. However, the use of some traits may threaten biodiversity and agrobiodiversity by limiting farmers' options to a few select varieties [ESAP Chapter 5; Global Chapters 3, 5, 6].

Breeding capacity is therefore of great importance to assessments of biotechnology in relation to sustainability and development goals [NAE Chapters 4, 6]. In developing countries, public plant breeding institutions are common but IP and globalization threaten them [Global Chapters 2, 6]. As privatization fuels a transfer of knowledge away from the commons, there is a contraction both in crop diversity and numbers of local breeding specialists. In many parts of the world women play this role, and thus a risk exists that privatization may lead to women losing economic resources and social standing as their plant breeding knowledge is appropriated. At the same time, entire communities run the risk of losing control of their food security [CWANA Chapter 1; Global Chapter 2].

Plant breeding activities differ between countries, so public investment in genetic improvement needs to be augmented by research units composed of local farming communities [Global Chapters 2, 6]. In addition, conflicts in priorities, that could endanger *in situ* conservation as a resource for breeding, arising from differences in IP protection philosophies need to be identified and resolved [Global Chapter 2]. For example, patent protection and forms of plant variety protection place a greater value on the role of breeders than that of local communities that maintain gene pools through *in situ* conservation [Global Chapter 2]. It will be important to find a new balance between exclusive access secured through IPR or other instruments and the need for local farmers and researchers to develop locally adapted varieties. It will be important to maintain a situation where innovation incentives achieved through IPR instruments and the need for local farmers and researchers to develop locally adapted varieties are mutually supportive. Patent systems, breeders' exemptions and farmers' privilege provisions may need further consideration here [Global Chapter 2]. An important early step may be to create effective local support for farmers. Support could come from, for example, farmer NGOs, where appropriate, to help develop local capacities, and advisers to farmer NGO's to guide their investments in local plant improvement. Participatory plant breeding, which incorporates TK, is a flexible strategy for generating new cultivars using different local varieties. It has the added advantage of empowering the local farmer and women [Global Chapter 2]. A number of *ad hoc* private initiatives for donating or co-developing IP are also appearing [Global Chapter 2], and more should be encouraged.

The decline in numbers of specialists in plant breeding, especially from the public sector, is a worrisome trend for maintaining and increasing global capacity for crop improvement [Global Chapter 6]. In addition, breeding supplemented with the use of MAS can speed up crop development, especially for simple traits [Global Chapter 3; NAE Chapter 6]. It may or may not also significantly accelerate the development of traits that depend on multiple genes [Global Chapter 6]. Provided that steps are taken to maintain local ownership and control of crop varieties, and to increase capacity in plant breeding, adaptive selection and breeding remain viable options for meeting development and sustainability goals [Global Chapter 6; NAE Chapter 6].

Gene flow

Regardless of how new varieties of crop plants are created, care needs to be taken when they are released because through gene flow they can become invasive or problem weeds, or the genes behind their desired agronomic traits may introgress into wild plants threatening local biodiversity [Global Chapter 5]. Gene flow may assist wild relatives and other crops to become more tolerant to a range of environmental conditions and thus further threaten sustainable production [Global Chapters 3, 6]. It is important to recognize that both biodiversity and crop diversity are important for sustainable agriculture. Gene flow is particularly relevant to transgenes both because they have tended thus far to be single genes or a few tightly linked genes in genomes, which means that they can be transmitted like any other simple trait through breeding (unlike some quantitative traits that require combinations of chromosomes to be inherited simultaneously), and because in the future some of the traits of most relevance to meeting development and sustainability goals are based on genes that adapt plants to new environments (e.g., drought and salt tolerance) [Global Chapter 5].

Transgene flow also creates potential liabilities [Global Chapter 6]. The liability is borne when the flow results in traditional, economic or environmental damage. For example, the flow of transgenes from pharmaceutical GM food crops to other food crops due to segregation failures could introduce both traditional and environmental damage. An important type of potential economic damage arises from the type of IPR instrument used to protect GM but not conventional and plants in some jurisdictions. The former are subject to IP protection that follows the gene rather than the trait, and is exempt from farmer's privilege provisions in some plant variety protection conventions [Global Chapter 6].

GMOs and chemical use

There is an active dispute over the evidence of adverse effects of GM crops on the environment [Global Chapter 3 vs. NAE Chapter 3]. That general dispute aside, as GM plants have been adopted mainly in high chemical input farming systems thus far [Global Chapter 3], the debate has focused on whether the concomitant changes in the amounts or types of some pesticides [Global Chapter 2; NAE Chapter 3] that were used in these systems prior to the development of commercial GM plants creates a net environmental benefit [Global Chapter 3]. Regardless of how this debate resolves, the benefits of current GM plants may not translate into all agroecosystems. For example, the benefits of reductions in use of other insecticides through the introduction of insecticide-producing (Bt) plants [NAE Chapter 3] seems to be primarily in chemically intensive agroecosystems such as North and South America and China [Global Chapter 3].

Livestock and aquaculture to increase food production and improve nutrition

Livestock, poultry and fish breeding have made substantial historical and current contributions to productivity [Global Chapters 3, 6, 7]. The key limitation to productivity increases in developing countries appears to be in adapting modern breeds to the local environment [CWANA Chapter 5; Global Chapter 3]. The same range of genomics and engineering options available to plants, theoretically, apply to livestock and fish [Global Chapters 3, 6; NAE Chapter 6]. In addition, livestock biotechnologies include artificial insemination, sire-testing, synchronization of estrus, embryo transfer and gamete and embryo cryopreservation, and new cloning techniques [see CWANA Chapter 5; Global Chapter 6; NAE Chapter 6 for a range of topics].

Biotechnology can contribute to livestock and aquaculture through the development of diagnostics and vaccines for infectious diseases [Global Chapter 6; NAE Chapter 6], transgenes for disease resistance [Global Chapter 3] and development of feeds that reduce nitrogen and phosphorous loads in waste [Global Chapter 3]. Breeding with enhanced growth characteristics or disease resistance is also made possible with MAS [Global Chapter 3; NAE Chapter 6]. As with plants, the difficulty with breeding animals is in bringing the different genes necessary for some traits together all at once in the offspring. Animals with desired traits might be more efficiently selected by using genomic maps to identify quantitative traits and gene x environment interactions.

There are currently no transgenic livestock animals in commercial production and none likely in the short term [Global Chapter 6]. Gene flow from GM fish also may be of significant concern and so GM fish would need to be closely monitored [CWANA Chapter 5; Global Chapter 3]. Assessing environmental impacts of GM fish is even more difficult than for GM plants, as even less is known about marine ecosystem than about terrestrial agroecosystems.

Ways Forward

Biotechnology must be considered in a holistic sense to capture its true contribution to AKST and achieving development and sustainability goals. On the one hand, this may be resisted because some biotechnologies, e.g., genetic engineering, are very controversial and the particular controversy can cause many to prematurely dismiss the value of all biotechnology in general. On the other hand, those who favor technologies that are most amenable to prevailing IP protections may resist broad definitions of biotechnology, because past contributions made by many individuals, institutions and societies might undermine the exclusivity of claims.

A problem-oriented approach to biotechnology R&D would focus investment on local priorities identified through participatory and transparent processes, and favor multifunctional solutions to local problems [Global Chapter 2]. This emphasis replaces a view where commercial drivers determine supply. The nature of the commercial organization is to secure the IP for products and methods development. IP law is designed to prevent the unauthorized use of IP rather than as an empowering right to develop products based on IP. Instead, there needs to be a renewed emphasis on public sector engagement in biotechnology. It is clearly realized that the private sector will not replace the public sector for producing biotechnologies that are used on smaller scales, maintaining broadly applicable research and development capacities, or achieving some goals for which there is no market [CWANA Chapter 5; Global Chapters 5, 8]. In saying this, an IP-motivated public engagement alone would miss the point, and the public sector must also have adequate resources and expertise to produce locally understood and relevant biotechnologies and products [CWANA Chapter 1].

A systematic redirection of AKST will include a rigorous rethinking of biotechnology, and especially modern biotechnology, in the decades to come. Effective long-term environmental and health monitoring and surveillance programs, and training and education of farmers are essential to identify emerging and comparative impacts on the environment and human health, and to take timely counter measures. No regional long-term environmental and health monitoring programs exist to date in the countries with the most concentrated GM crop production [Global Chapter 3]. Hence, long-term data on environmental implications of GM crop production are at best deductive or simply missing and speculative.

While climate change and population growth could collude to overwhelm the Earth's latent potential to grow food and bio-materials that sustain human life and well being, both forces might be offset by smarter agriculture. Present cultivation methods are energy intensive and environmentally taxing, characteristics that in time both exacerbate demand for limited resources and damage long term productivity. Agroecosystems that both improve productivity and replenish ecosystem services behind the supply chain are desperately needed. No particular actor has all the answers or all the possible tools to achieve a global solution. Genetically modified plants and GM fish may have a sustainable contribution to make in some environments just as ecological agriculture might be a superior approach to achieving a higher sustainable level of agricultural productivity.

Climate Change

Writing team: Gordana Kranjac-Berisavljevic (Ghana), Balgis Osman-Elasha (Sudan), Wahida Patwa Shah (Kenya), John M.R. Stone (Canada)

Why is climate change important to achieving development and sustainability goals? The threat of climate change contains the potential for irreversible damage to the natural resource base on which agriculture depends and hence poses a grave threat to development. In addition, climate changes are taking place simultaneously with increasing demands for food, feed, fiber and fuel [ESAP Chapter 4; NAE Chapter 3]. Addressing these issues will require a wide range of adaptation and emission reduction measures.

The climate change issue presents decision makers with a set of formidable challenges not the least of these is the inherent complexity of the climate system [CWANA Chapter 1; ESAP Chapter 4; LAC Chapter 3; NAE Chapter 3]. These complexities include the long time lags between greenhouse gas[11] emissions and effects, the global scope of the problem but wide regional variations, the need to consider multiple greenhouse gases and aerosols, and the carbon cycle, which is important for converting emissions into atmospheric concentrations. Another significant challenge is the rapidity of the changes in the climate that have occurred or will occur [NAE Chapter 3].

Dependency of agriculture on climate. Agricultural production depends on the provision of essential natural ecosystems inputs such as adequate water quantity and quality, soil nutrients, biodiversity and atmospheric carbon dioxide to deliver food, fiber, fuel and commodities for human use and consumption. The ecosystem services that provide these inputs are affected, both directly and indirectly, by climate change [CWANA Chapter 1; ESAP Chapters 2, 4; Global Chapter 1; SSA Chapter 4]. Climate change, for example, can affect the agrobiodiversity necessary for crop, tree and livestock improvement, pest control and soil nutrient cycling.

Agricultural production has always been affected by

---

[11] Greenhouse gases and clouds in the atmosphere absorb the majority of the long-wave radiation emitted by the Earth's surface, modifying the radiation balance and, hence, the climate of the Earth. The primary greenhouse gases are of both, natural and anthropogenic origin, including water vapour, carbon dioxide [CO_2], methane [CH_4] nitrous oxide [N_2O] and ozone [O_3], while halocarbons and other chlorine- and bromine-containing substances are entirely anthropogenic.

natural climate variability and extreme climate events have caused significant damage to agriculture and livelihoods resulting in food insecurity and poverty among rural communities [CWANA Chapter 3; ESAP Chapter 4; LAC Chapter 3; NAE Chapters 2,3; SSA Chapter 1]. Throughout human history people all over the world have learned to adapt to such climate variability and extreme events. However, experience with adaptive measures differs widely among regions, countries and continents, as do the risks involved [NAE Chapter 3]. This Assessment provides many example of climate change's effects on food production, agroforestry, animal production systems, fisheries and forestry [CWANA Chapter 1; ESAP Chapters 2, 4; LAC Chapter 3; NAE Chapters 1, 3; SSA Chapter 4]. Poor, forest dependent people and small-scale fishers who lack mobility and livelihood alternatives suffer disproportionately from climatic variability. The El Nino-Southern Oscillation (ENSO) phenomenon, associated with massive fluctuations in the marine ecosystems of the western coast of South America, adversely affects fishing and has led to devastating socioeconomic tolls on the communities that depend on this activity [LAC Chapter 1] Access to training, education, credit, technologies and other agricultural resources affects the ability of women in particular to cope with climate change-induced stresses.

Dependency of climate on agriculture. The relationship between climate change and agriculture (crops, livestock and forestry) is not a one-way street. [Global Chapter 1; NAE Chapter 2]. Agriculture contributes to climate change in several major ways including:

- Land conversion and plowing releases large amounts of stored carbon as CO_2 from vegetation and soils. About 50% of the world's surface land area has been converted to land for grazing and crop cultivation resulting in a loss of more than half of the world's forests. Deforestation and forest degradation releases carbon through the decomposition of aboveground biomass and peat fires and decay of drained peat soils.
- Carbon dioxide (CO_2) and particulate matter are emitted from fossil fuels used to power farm machinery, irrigation pumps, and for drying grain, etc., as well as fertilizer and pesticide production [NAE Chapter 2].
- Nitrogen fertilizer applications and manure applications as well as decomposition of agricultural wastes results in emissions of nitrous oxide (N_2O).
- Methane (CH_4) is released through livestock digestive processes and rice production.

- Altered radiative fluxes and evaporation from newly bare soils [Global Chapter 3].
- Increased geographical distance between producer and consumer, together with regional agricultural specialization, has resulted in greater energy use for transportation.

Overall, agriculture (cropping and livestock) contributes 13.5 % of global greenhouse gas emissions mostly through emissions of methane and nitrous oxide (about 47% and 58% of total anthropogenic emissions of CH_4 and N_2O, respectively). However reports from other estimate the emissions from livestock alone to account for 18% of total emissions. This figure includes the entire commodity chain for livestock. Land use, land use change and forestry contribute another 17.4% mostly as carbon dioxide. Most of greenhouse gas emissions are from land use changes and soil management (40%), enteric fermentation (27%), and rice cultivation (10%). As diets change and there is more demand for meat, there is the potential for increased GHG emissions from agriculture. The relative contribution varies by region; in NAE it is estimated to be in the range of 7-20% [Global Chapter 1; NAE Chapter 2]. The highest emissions of greenhouse gases from agriculture are generally associated with the most intensive farming systems. Sub-Saharan Africa, on rainfed agriculture, contributes the least in terms of GHG emissions and yet it is among the most vulnerable regions to the impacts of climate change [NAE Chapter 3; SSA Chapter 1] due to multiple stresses, including the heavy reliance on rain fed agriculture, poverty, weak institutional structures and low adaptive capacity.

Changes in land use have negatively affected the net ability of ecosystems to sequester carbon from the atmosphere. For instance, the carbon rich grasslands and forests in temperate zones have been replaced by crops with much lower capacity to sequester carbon. Despite a slow increase in forests in the northern hemisphere, the overall benefits in terms of carbon sequestration are being lost due to increased deforestation in the tropics. There are however complex tradeoffs, for example, when forest is replaced by oil palm which will capture carbon but reduce biodiversity. Climate change is also likely to affect the carbon cycle and some vulnerable natural pools of carbon could turn into sources, e.g., loss of peatlands. [Global Chapter 1; NAE Chapter 3].

Observed climate change and impacts. Overall, longer and more intense droughts have been observed since the 1970s, particularly in the tropics and sub-tropics. Extreme events such as floods, droughts and tropical cyclones are now more intense than before. Throughout NAE there have been significant increases in serious forest fires, in part due to climate change, dense biomass and more human access into remote areas. The thermal growing season has lengthened by about 10 days.

Poor, forest dependent people and small-scale fishers who lack mobility and livelihood alternatives suffer disproportionately from climatic variability. The El Niño-Southern Oscillation (ENSO) phenomenon, associated with massive fluctuations in the marine ecosystems of the western coast of South America, adversely affects fishing, and has lead to devastating socioeconomic tolls on the communities that depend on this activity [LAC Chapter 1].

Future climate change and projected impacts. Increased growth and yield rates due to higher levels of carbon dioxide and temperatures could result in longer growing seasons. For example, in mid- to high-latitude regions, according to the Intergovernmental Panel on Climate Change's

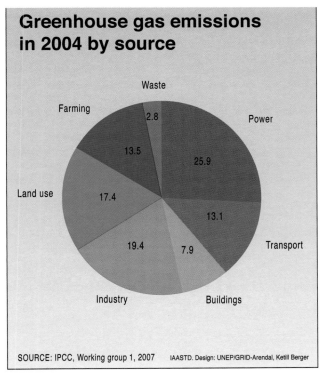

Figure SR-CC1a. *Greenhouse gas (GHG) emissions, 2004.*

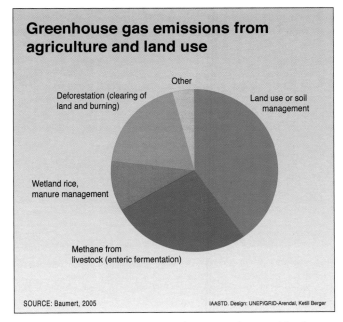

Figure SR-CC1b. *GHG emissions from agriculture and land use.*

(IPCC) Fourth Assessment Report moderate local increases in temperature (1-2°C) can have small beneficial impacts on crop yields. However, in low-latitude regions, even such moderate temperature increases are likely to have negative yield impacts for major cereals. *Some negative impacts are already visible, especially in developing countries.* [ESAP Chapter 2; Global Chapter 5; NAE Chapter 3]. Further warming will have increasingly negative impacts, particularly affecting production in food insecure regions. Warming in NAE will lead to a northward expansion of suitable cropping areas as well as a reduction of the growing period of crops such as cereals, but results, on the whole, project the potential for global food production to increase with increases in local average temperature over a range of 1 to 3°C, and above this range to decrease.

From an ecosystem perspective, the rate of change can be more important. By 2030, temperature increases of more than 0.2 C° per decade are projected. Rates in excess of this are considered by some experts to be dangerous, although our current understanding is still uncertain [Global Chapter 5].

Although the state of knowledge of precipitation changes is currently insufficient for confidence in the details, we expect that for many crops water scarcity will increasingly constrain production. Climate change will require a new look at water storage to cope with the impacts of changes in total amounts of precipitation and increased

rates of evapotranspiration, shifts in ratios between snowfall and rainfall and the timing of water availability, and with the reduction of water stored in mountain glaciers. Many climate impact studies project global water problems in the near future unless appropriate action is taken to improve water management and increase water use efficiency. Projections suggest that by 2050 internal renewable water is estimated to increase in some developed countries, but is expected to decrease in most developing countries [Global Chapter 5].

Climate change will increase heat and drought stress in many of the current breadbaskets in China, India, and the United States and even more so in the already stressed areas of sub-Saharan Africa. Rainfed agriculture, especially of rice and wheat in the ESAP, is likely to be vulnerable. For example, rainfed rice yield could be reduced by 5-12% in China for a 2°C rise in temperature. [ESAP Chapter 4; Global Chapter 6; NAE Chapter 3].

Most climate models indicate a strengthening of the summer monsoon and increased rainfall in Asia, but in semiarid areas in Africa the absolute amount of rain may decline, and seasonal and inter-annual variation increase. Reductions in the duration or changes in timing of the onset of seasonal floods will affect the scheduling and extent of the cropping and growing seasons, which may in turn have large impacts on livelihoods and production systems. For example, droughts occurring in the monsoon period se-

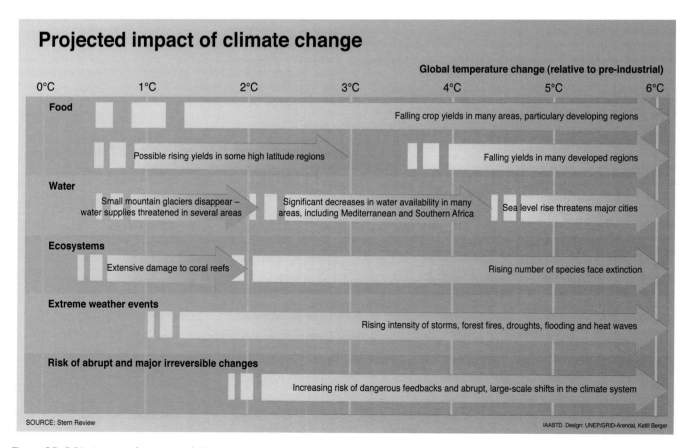

Figure SR-CC2. *Projected impacts of climate change.*

verely affect rice crop production in ESAP [ESAP Chapter 4; Global Chapter 5].

Extreme climate events are expected to increase in frequency and severity and all regions will likely be affected by the increase in floods, droughts, heat waves, tropical cyclones and other extreme events with significant consequences for food and forestry production, and food insecurity. This was demonstrated during the summer 2003 European heat wave that was accompanied by drought and reduced maize yields by 20 percent. There is likely to be an increase in incidence and severity of forest fires in next decades, partly as a result of climate change [NAE Chapter 2].

Climate change is expected to threaten livestock holders in numerous ways: animals are very sensitive to heat stress; they require a reliable resource of water and pasture is very sensitive to drought. In addition, infectious and vector-borne animal diseases will continue to become increasingly frequent worldwide [Global Chapter 3].

The effects of climate change on crop and tree yields, fisheries, forestry and livestock vary greatly by region [Global Chapter 1; SSA Chapter 4] and climate scenarios project that local biomes and terrestrial ecosystems will change. Although climate projections cannot tell us exactly what and where the changes will be and when they will be experienced, it is known that climate change will affect regional patterns of temperature and precipitation.

Global climate change is expected to alter marine and freshwater ecosystems and habitats. Rising sea levels will alter coastal habitats and their future productivity, threatening some of the most productive fishing areas in the world. Changes in ocean temperatures will alter ocean currents and the distribution and ranges of marine animals, including fish populations. Rising atmospheric CO_2 will lead to acidification of ocean waters and disrupt the ability of animals (such as corals, mollusks, plankton) to secrete calcareous skeletons, thus reducing their role in critical ecosystems and food webs [Global Chapter 6; SSA Chapter 4]. Sea level rise could lead to saltwater intrusion causing a reduction in agricultural productivity in some coastal areas [ESAP Chapters 2, 4; Global Chapter 1; NAE Chapter 3; SSA Chapter 3]. It is expected that climate change will lead to significant reductions in the diversity fish species with important changes in abundance and distribution of fresh water fish stocks such as in rivers and lakes in SSA.

Climate change is affecting and will affect the geographic range and incidence of many human, animal, and plant pests, disease vectors and wide variety of invasive species that will inhabit new ecological niches, [ESAP Chapter 3; Global Chapters 1, 5, 6, 7]. These anticipated changes may have a negative impact on agricultural activities through their effect on the health of farmers and ecosystems, particularly in developing countries. For example, an increase in temperature and precipitation is projected to expand the range of vector-transmitted diseases making it possible for these diseases to become established outside limits of their current range, and at higher elevations [LAC Chapter 1]. In addition, increased irrigation as an adaptive response to better control water scarcity due to climate change may increase incidences of malaria [Global Chapter 5] and other water-related diseases.

Pests and diseases are strongly influenced by seasonal weather patterns and changes in climate. Established pests may become more prevalent due to favorable conditions that include higher winter temperatures (thus reduced winter-kill) and more rainfall. New pest introductions alter pest/predator/parasite population dynamics through changes in growth and developmental rates, the number of generations produced per year, the severity and density of populations, the pest virulence to a host plant, or the susceptibility of the host to the pest. Changing weather patterns also increase crop vulnerability to pests, weeds and invasive plants, thus decreasing yields and increasing pesticide applications [Global Chapter 3]. Increased temperatures are likely to facilitate range expansion of highly damaging weeds, which are currently limited by cool temperatures [Global Chapters 3, 6].

Climate simulation models indicate substantial future increases in soil erosion. Tropical soils with low organic matter are expected to experience the greatest impact of erosion on crop productivity. Desertification will be exacerbated by reductions in average annual rainfall and increased evapotranspiration especially in soils that have low levels of biological activity, organic matter and aggregate stability. [CWANA Chapter 1; Global Chapter 6] In addition, continued migration to urban areas of younger segments of the population can lead to agricultural land degradation thus exacerbating the effects of climate change, as those left on the land are mostly old and the vulnerable.

There is a serious potential for future conflict, and possible violent clashes over habitable land and natural resources, such as freshwater, as a result of climate change, which could seriously impede food security and poverty reduction. An estimated 25 million people per year already flee from weather-related disasters; global warming is projected to increase this number to some 200 million before 2050, with semiarid ecosystems expected to be the most vulnerable to impacts from climate change refugees [Global Chapter 6]. In addition, climate change combined with other socioeconomic stresses could alter the regional distribution of hunger and malnutrition, with large negative effects on sub-Saharan Africa.

Options for Action

The IPCC concluded that "warming of the climate system is now unequivocal" and that "most of the observed increase in globally averaged temperatures since the mid-20th century is *very likely* due to the observed increase in anthropogenic greenhouse gas concentrations." With these strong conclusions the focus should now shift from defining the threat to seeking solutions.

In considering responses to the threat of climate change there are important policy considerations. Tackling the root cause of the problem, which is the emissions of greenhouse gases into the atmosphere, requires a global approach. The earlier and stronger the cuts in emissions, the quicker concentrations will approach stabilization. While emission reduction measures clearly are essential, further changes in the climate are now inevitable and thus adaptation becomes imperative. Climate change is not simply an environmental issue but can also be framed in terms of other issues such as sustainable development and security. Actions directed at addressing climate change and efforts to promote sus-

Table CC1. **Multiple stressors in small-scale agriculture.**

Multiple stressors in small-scale agriculture

Stressors	Source
Population increase driving fragmentation of landholding	Various
Environmental degradation stemming variously from population, poverty, ill-defined property rights	Grimble et al., 2002
Regionalised and globalised markets, and regulatory regimes, increasingly concerned with issues of food quality and food safety	Reardon et al., 2003
Market failures interrupt input supply following withdrawal of government intervention	Kherallah et al., 2002
Continued protectionist agricultural policies in developed countries, and continued declines and unpredictability in the world prices of many major agricultural commodities of developing countries	Lipton, 2004, Various
Human immunodeficiency virus (HIV) and/or acquired immunodeficiency syndrome (AIDS) pandemic, particularly in Southern Africa, attacking agriculture through the deaths of working age adults, which diverts labor away from farming, erodes household assets, disrupts knowledge transfer and reduces the capacity of agricultural service providers	Barnet and Whiteside, 2002
For pastoralists, encroachment on grazing lands and failure to maintain traditional natural resource management	Blench, 2001
State fragility and armed conflict in some regions	Various

SOURCE: IPCC, 2007.

tainable development share some important common goals and determinants such as, for example, equitable access to resources, appropriate technologies and decision-support mechanisms to cope with risks. Furthermore, decisions on climate change are usually made in the context of other environmental, social and economic stresses.

There is a need to develop agricultural policies that both reduce emissions and allow adaptation to climate change that are closer to carbon-neutral, minimize trace gas emissions and reduce natural capital degradation [Global Chapter 4]. Important questions include how emissions from agriculture and forestry can be effectively reduced, how to produce food with greater input efficiency and less GHG emissions, how agriculture, agroforestry and forestry can best adapt under given local conditions, and what role biofuels can play—and, finally, what are the implications of these challenges on requirements for AKST [NAE Chapter 3]. More efforts will be required to develop new knowledge and technologies, especially for energy-efficient farming systems, as well as more comprehensive cost-benefit analysis than these now available [Global Chapter 3]. Interconnected issues, such as the effects of land use changes on biodiversity and on land degradation, need to be addressed in order to exploit synergies between the goals of UN conventions on biodiversity and desertification and climate change.

Adaptation and mitigation are complementary strategies to reduce impacts. The effects of reduced emissions in avoiding impacts by slowing the rate of temperature increase will not emerge for several decades due to the inertia of the climate system. Adaptation, therefore, will be important in coping with early impacts. Specifically, adaptation will be necessary to meet the challenge of impacts on agriculture

to which we are already committed in the near term as well as for the long term, where the risk of unmitigated climate change impacts could exceed the adaptive capacity of existing agricultural systems.

Some "win-win" mitigation opportunities have already been identified. These include land use approaches such as lower rates of agricultural expansion into natural habitats; afforestation, reforestation, agroforestry and restoration of underutilized or degraded land; land use options such as carbon sequestration in agricultural soils, appropriate application of nitrogenous inputs; and effective manure management and use of feed that increases livestock digestive efficiency.

Policy options covering regulations and investment opportunities include financial incentives to maintain and increase forest area through reduced deforestation and degradation and improved management. Those policy options that enhance the production of renewable energy sources could be particularly effective. Any post-2012 regime has to be more inclusive of all agricultural such as reduced emission from reforestation and degradation activities to take full advantage of the opportunities offered by agriculture and forestry sectors [Global Chapter 6].

Local, national and regional agricultural development regulatory frameworks will have to take into account tradeoffs between the need for promoting higher yields and the need for the maintenance and enhancement of environmental services that support agriculture [SSA Chapter 4].

Adaptation options. Two types of adaptation have been recognized: autonomous and planned adaptation. Autonomous adaptation does not constitute a conscious response to climatic stimuli but is triggered by ecological changes in natural systems and by market or welfare changes in human systems. Planned adaptation is the result of a deliberate policy decision, based on an awareness that conditions have changed or are about to change and that action is required to return to, maintain, or achieve a desired state. It could also take place at the community level, triggered by knowledge of the future impacts of climate change and realization that extreme events experienced in the past are likely to be repeated in the future. The first means the implementation of existing knowledge and technology in response to the changes experienced, while the latter means the increased adaptive capacity by improving or changing institutions and policies, and investments in new technologies and infrastructure to enable effective adaptation activities.

Many autonomous adaptation options are largely extensions or intensifications of existing risk-management or production-enhancement activities. These include:
- Changing varieties/species to fit more appropriately to the changing thermal and/or hydrological conditions;
- Changing timing of irrigation and adjusting nutrient management;
- Applying water-conserving technologies and promoting agrobiodiversity for increased resilience of the agricultural systems; and
- Altering timing or location of cropping activities and the diversification of agriculture [Global Chapter 6].

Planned adaptations include specific policies are aiming at reducing poverty and increasing livelihood security, provision of infrastructure that supports/enables integrated spatial planning and the generation and dissemination of new knowledge and technologies and management practices tailored to anticipated changes [NAE Chapter 3]. It is important to note that policy-based adaptations to climate change will interact with, depend on or perhaps even be just a subset of policies on natural resource management, human and animal health, governance and political rights, among many others. These represent examples of the "mainstreaming" of climate change adaptation into policies intended to enhance broad resilience.

The extent to which development and sustainability goals will be affected by climate change depends on how well communities are able to cope with current climate change and variability, as well as to other stresses such as land degradation, poverty, lack of economic diversification, institutional stability and conflict [Global Chapter 6]. Industrialized world agriculture, generally situated at high latitudes and possessing economies of scale, good access to information, technology and insurance programs, as well as favorable terms of global trade, is positioned relatively well to adapt to climate change. By contrast small-scale rain-fed production systems in semiarid and subhumid zones, which continuously face significant seasonal and inter-annual climate variability, are characterized by poor adaptive capacity due to the marginal nature of the production environment and the constraining effects of poverty and land degradation [Global Chapter 6]. Sub-Saharan Africa and CWANA are especially vulnerable regions [CWANA Chapter 1; SSA Chapter 1]. The resilience of dry-land ecosystems to deficits in moisture, temperature extremes and salinity is still inadequately understood.

The effectiveness of AKST's adaptation efforts is likely to vary significantly between and within regions, depending on exposure to climate impacts and adaptive capacity, the latter depending very much on economic diversification and wealth and institutional capacity. The viability of traditional actions taken by people to lessen the impacts of climate change in arid and semi arid regions depends on the ability to anticipate hazard patterns, which are getting increasingly erratic. Early detection and warning using novel GIS-based methodologies such as those employed by the Conflict Early Warning and Response Network (CEWARN) and the Global Public Health Information Network (G-PHIN) could play a useful role.

Bringing climate prediction to bear on the needs of agriculture requires increasing observational networks in the most vulnerable regions, further improvements in forecast accuracy, integrating seasonal prediction with information at shorter and longer time scales, embedding crop models within climate models, enhanced use of remote sensing, integration into agricultural risk management, enhanced stakeholder participation, and commodity trade and storage applications [Global Chapter 6].

Mitigation options. A number of options, technologies and techniques to reduce or off-set the emissions of GHGs already exist and could:
- Lower levels of methane or nitrous oxide through increasing the efficiency of livestock production, improving animals' diets and using feed additives to increase food conversion efficiency, reducing enteric fermentation and consequent methane emissions, aerating manure before composting and recycling agricultural and forestry residues to produce biofuels.
- Lower nitrous oxides emissions through matching manure and fertilizer application to crop needs and optimizing nitrogen up-take efficiently by controlling the application rates, method and timing.
- Reduce emissions from deforestation and forest degradation, including policy measures to address drivers of deforestation, improve forest management, forest law enforcement, forest fire management, improve silvicultural practices and promote afforestation and reforestation to increase carbon storage in forests [Global Chapters 1, 3, 5, 6; SSA Chapter 3]
- Improve the soil carbon retention by promoting biodiversity as a tool for climate mitigation and adaptation and enhance the management of residues, using zero/reduced tillage, including legumes in crop rotation, reducing the fallow periods and converting marginal lands into woodlots. [Global Chapters 1, 3, 5, 6; SSA Chapter 3]
- Support low-input farming agriculture that relies on renewable sources of energy.

It is important that efforts aimed at addressing emissions reductions mitigation from agriculture carefully consider all potential GHG emissions. For example, efforts to reduce CH_4 emissions in rice could lead to greater N_2O emissions through changes in soil N dynamics. Similarly, conservation tillage for soil carbon sequestration can result in elevated N_2O emissions through increased agrochemicals use and accelerated denitrification in soils [Global Chapter 6].

In addition, policy options regulations and investment opportunities that include financial incentives to increase forest area, reduce deforestation and maintain and manage forests, enhance the production of renewable energy sources could be particularly effective. However, some challenges may arise in developing countries which lack sufficient investment capital and have unresolved land tenure issues [Global Chapters 1, 3, 5; SSA Chapter 3].

Climate change regimes. The Kyoto Protocol currently represents the highest level of international consensus around the need to address climate change. Questions have been raised regarding its effectiveness in reducing global emissions to avoid dangerous climate change. It is clear that the Kyoto Protocol is a first step, one that demonstrates political will and allows for some policy experimentation, and that deeper cuts and additional de-carbonization strategies are needed. Mitigation options employing the agricultural sectors are not well covered under the Protocol. In this regard a much more comprehensive future looking agreement is needed if we want to take full advantage of the opportunities offered by agriculture and forestry sectors.

Achieving this could be accomplished through a negotiated global long-term (30-50 years), comprehensive and equitable regulatory framework with differentiated responsibilities and intermediate targets to reduce the GHG emissions. Within such a framework a modified Clean Development Mechanism (CDM) with a comprehensive set of eligible agricultural mitigation activities, including afforestation and reforestation; avoided deforestation, using a national sectoral approach rather than a project approach to minimize issues of leakage, thus allowing for policy interventions; and a wide range of agricultural practices including organic agriculture and conservation tillage could help meet the development and sustainability goals. Other approaches could include reduced agricultural subsidies that promote GHG emissions and mechanisms to encourage and support adaptation, particularly in vulnerable regions, such as the tropics and sub-tropics.

Human Health

Writing Team: Kristie L. Ebi (USA), Rose R. Kingamkono (Tanzania), Karen Lock (UK), Yalem Mekonnen (Ethiopia)

Inter-linkages between health, nutrition, agriculture, and AKST affect the ability of individuals, communities, and nations to reach sustainability goals. These interlinkages take place within a context of other, multiple stressors that affect population health. Intake of food of insufficient quantity, quality, and variety can result in ill-health. Poor health in adults and children leads to reduced economic productivity. Malnutrition and recurrent infections in childhood impair physical growth and mental development, thus lowering economic productivity in adulthood [Global Chapters 1, 3, 6; SSA]. Lowered immunity associated with undernutrition makes individuals more susceptible to a range of diseases, including HIV/AIDS, and can make treatment and recovery more difficult [CWANA; ESAP; Global Chapters 2, 3, 5; LAC; SSA]. Improving health by controlling a range of infectious and chronic diseases can increase the effectiveness and productivity of food systems and AKST.

Agriculture has generally not had an explicit goal of improving human health. Appropriate application of AKST can improve dietary quantity and quality and overall population health; Examples include appropriate crop diversification approaches; the use of fertilizers, such as zinc, selenium, and iodine, on soils low in these essential human nutrients; and development of agroecosystem farming approaches designed to improve human, animal, and soil health [Global Chapters 2, 3, 5, 6, 8].

Agriculture can inadvertently affect health through the emergence of infectious diseases (approximately 75% of emerging diseases are zoonotic—transmitted between animals and humans) [Global Chapters 3, 5, 6, 9; NAE Chapters 1, 4; SSA Chapter 3]. Furthermore, agriculture is one of the three most dangerous occupations [with mining and construction] in terms of deaths, accidents, exposures, and occupationally related ill-health [Global Chapter 3]. Consumers are increasingly worried about increased risk of ill-health resulting from exposure to pesticides and other agrichemicals, antibiotics and growth hormones, additives introduced during food-processing, and foodborne pathogens [CWANA Chapter 5; ESAP Chapters 2, 3, 5; Global Chapters 2, 3, 5, 6, 8; LAC Chapter 1; NAE Chapter 2; SSA Chapters 2, 3].

Current Status and Trends

Interrelationship between poor health and agriculture. Vulnerable populations, particularly in rural communities, are typically exposed to multiple and interacting health risks associated with agriculture, including poor nutrition, food safety, and occupational and environmental health risks. This often results in a significant cumulative burden of ill health.

Poor health in turn impacts on multiple agricultural functions and outputs. High prevalence rates of malnutrition and infectious and chronic diseases decrease productivity through labor shortages, the need to change the type of crops grown, and the need to reduce the total area of land under cultivation. Poor health also impacts on farmers' ability to innovate and develop new farming systems. Ill health among families of producers can impact on production through absenteeism to provide health and other care, and the loss of household income or other outputs of agricultural work [CWANA; ESAP; Global Chapter 3; LAC; NAE; SSA]. This is particularly important for women who are often both the primary producers and primary carers [see Women in Agriculture theme]. Reduced life expectancy results in loss of local agricultural knowledge and reduced capacity, especially with respect to uptake of AKST. In developing countries these issues are clearly illustrated by the impact of HIV-AIDS, malaria and malnutrition [CWANA; ESAP; Global Chapters 1, 3; LAC; SSA].

Malnutrition. Worldwide, ill health due to poor nutrition results from under-nutrition over-nutrition, and imbalanced food intake leading to obesity [CWANA; ESAP; Global Chapters 1, 2, 3; LAC; NAE Chapter 2; SSA Chapter 2]. Individual risk factors for under-nutrition include insufficient macro- or micronutrient dietary intake; depletion of body nutrients due to infections; and increased nutrient requirements during childhood, adolescence, pregnancy, and high physical activity such as manual labor. Malnutrition in many countries and regions continues to result from food insecurity due to multiple causes including loss of land, economic and political instability, war, and extreme climate events [Global Chapters 1, 3; SSA Chapter 2].

Over the past 40 years, there have been significant increases in global food production and supply that has surpassed population growth in many countries [Global Chapters 1, 2, 3]. During this period, global under-nutrition declined but still remains a major public health problem, estimated to contribute to over 15% of the total global burden of disease in 2000, with high variability in the extent of the problem between and within countries. Between 1981 and 2003, 97 developing and 27 transitional countries had a poor Global Hunger Index [GHI].[12] [Global Chapter 2] In Africa, particularly sub-Saharan Africa, chronic food shortages meant that trends in malnutrition continued or worsened over the past decades [SSA Chapters 1, 2, 3].

Although the world food system provides an adequate supply of protein and energy for over 85% of people, only two-thirds have access to sufficient dietary micronutrients [Global Chapters 1, 3]. The supply of many nutrients in the diets of the poor has decreased due to a reduction in diet diversity resulting from increased monoculture of staple food crops (rice, wheat, and maize) and the loss of a range of nutrient dense food crops from local food systems. Micronutrient deficiencies lower productivity, in both developed and developing countries, due to compromised health and impaired cognition. [CWANA; ESAP; Global Chapters 1, 2, 3; LAC; SSA].

Dietary-related chronic diseases. The success of AKST policies and practices in increasing production and in new mechanisms for processing foods have facilitated increasing rates of worldwide obesity and chronic disease through negative changes in dietary quality [Global Chapters 1, 2, 3, 6; NAE]. Worldwide changes in food systems have resulted in overall reductions in dietary diversity, with low population consumption of fruits and vegetables and high intakes of fats, meat, sugar and salt [Global Chapters 1, 2, 3; NAE]. Poor diet throughout the life course is a major risk factor for chronic diseases (including heart disease, stroke, diabetes and cancer) [Global Chapters 1, 3, 6; NAE Chapter 2] that comprise the largest proportion of global deaths. Together with environmental factors such as rapid urbanization which result in increased sedentary lifestyles (motorized transport, etc.), dietary changes contribute to continuing global increases in chronic diseases, overweight, and obesity affecting both rich and poor in developed and developing countries. The most dramatic rises in obesity are now occurring in low- and middle-income countries [Global Chapters 1, 2, 3; NAE Chapter 2]. These nutrition-related chronic diseases coexist with under-nutrition in many countries causing a greater disease burden in lower income countries [Global Chapters 1, 2, 3]. Unless action is taken to reduce these trends, all countries will see an increase in the economic burden due to loss of productivity, increased health care and social welfare costs that are already seen in developed countries [Global Chapter 3; NAE]. Many national and international actors have been slow to understand and adapt their policies to address these worldwide changes occurring in diet, nutrition, and their health impacts [Global Chapters 1, 2, 3; NAE Chapter 2].

Policies, regimes and consumer demands have tended to increase production (especially in the US and Europe) of, and processing incentives for, foodstuffs that are risk factors for chronic disease (high fat dairy, meat, etc.) [Global

Chapter 3; NAE Chapter 2]. AKST has focused on adding financial value to basic foodstuffs (e.g., using potatoes to produce a wide range of snack foods). This has resulted in cheap, processed food products with low nutrient density (high in fat, refined sugars and salt), and that have a long shelf life. Increased consumption of these food products that are replacing more varied, traditional diets, is contributing to increased rates of obesity and diet-related chronic disease worldwide. This has been exacerbated by the significant role of huge advertising budgets spent on unhealthy foods. There are a few examples of agricultural food policies that have been developed due to population health concerns; e.g., formation of the EU common agricultural policy whose original objectives included food security. In contrast, recent national and international agricultural trade policies/ regimes have not addressed the changing global health challenges and do not have explicit public heath goals.

Food safety. Although subject to controls and standards, globalization of the food supply, accompanied by concentration of food distribution and processing companies, and growing consumer awareness, increase the need for effective, coordinated, and proactive national food safety systems [CWANA Chapter 5; ESAP Chapters 2, 3, 5; Global Chapters 2, 3, 5, 6, 7, 8; LAC Chapter 1; NAE Chapters 1, 2; SSA Chapters 2, 3]. Issues include accountability and lack of vertical integration between consumers and producers. A food hazard is a biological, chemical, or physical contaminant, or an agent that affects bioavailability of nutrients. Food safety hazards may be introduced anywhere along the food chain with many hazards resulting from inputs into production and handling of commodities [Global Chapter 2]. As food passes through a multitude of food handlers and middlemen over extended period of time through the food production, processing, storage, and distribution chain, control has become difficult, increasing the risks of exposing food to contamination or adulteration. Concerns that could be addressed by AKST include heavy metals, pesticides, safe use of biofertilizers, the use of hormones and antibiotics in meat production, large-scale livestock farming and the use of various additives in food-processing industries. In general, developed countries, despite long food chains, guarantee a high level of consumer protection of imported and domestic food supplies; the capacity and legislative frameworks of public health systems quickly identify and control disease outbreaks. In developing countries, safety concerns are compounded by poverty; inadequate infrastructure for enforcement of food control systems; inadequate social services and structures (potable water, health, education, transportation); population growth; high incidence and prevalence of communicable diseases including HIV/AIDS; and trade pressure [CWANA Chapter 5; ESAP Chapters 2, 3, 5; LAC Chapter 1; NAE Chapters 1, 2; SSA Chapters 2, 3].

AKST control of food contamination creates social and economic burdens on communities and their health systems through market rejection costs of contaminated commodities causing export market losses, the need for sampling and testing, costs to food processors and consumers, and associated health costs [Global Chapters 2, 5, 7, 8]. The incidence of foodborne illnesses caused by pathogenic biological food

---

[12] GHI captures three equal weighted indicators of hunger: insufficient availability of food [the proportion of people who are food energy deficient]; short fall in nutritional status of children [prevalence of underweight for <5 years old children] and child mortality [<5 years old mortality rate] which are attributable to undernutrition.

contaminants, including bacteria, fungi, viruses, or parasites, has increased significantly over the past few decades [Global Chapters 1, 3, 5]. In developing countries, foodborne diseases can cause and/or exacerbate malnutrition. Together, these cause an estimated 12 to 13 million child deaths; survivors are often left with impaired physical and/or mental development that limits their ability to reach their full potential [Global Chapter 1].

There is increasing public concern over new AKST technologies, including GMOs and food irradiation. There is no clear scientific consensus whether these technologies affect population health. Significant knowledge gaps limit the assessment of the human health risks of GMOs. Food irradiation although useful in reducing the risk of microbial foodborne illness, could pose dangers to consumers, workers, and the environment [Global Chapters 1, 2, 5].

Occupational impacts on health. Worldwide, agriculture accounts for at least 170,000 occupational deaths each year. This number accounts for half of all fatal accidents worldwide and is likely an underestimate as most injuries are underreported in developing countries [Global Chapter 3]. Machinery and equipment, such as tractors and harvesters, account for the highest rates of injury and death [Global Chapters 1, 3]. Other health hazards include agrichemicals; transmissible animal diseases; toxic or allergenic agents; and noise, vibration, and ergonomic hazards (related to heavy loads, repetitive work, and inadequate equipment). Exposure to pesticides and other agrichemicals constitutes a major hazard to occupational health (and also wider community environmental health), with poisoning leading to acute, sub-acute, and chronic adverse health impacts (e.g., neurotoxicity, respiratory, and reproductive impacts), particularly among vulnerable populations, and to death including suicide [Global Chapters 2, 3; SSA]. The WHO has estimated that between 2 to 5 million cases of pesticide poisoning occur each year, resulting in approximately 220,000 deaths. This figure is widely recognized to be an underestimate based on empirical research [Global Chapters 2, 3, 7]. Even when used according to manufacturers specifications, following best practice and all protective measures, pesticide exposure cannot be avoided entirely and therefore some element of risk will remain particularly with highly toxic products. This is particularly relevant for developing countries, where conditions of poverty and lack of effective controls on hazardous compounds are the norm [Global Chapters 1, 2, 3]. In less developed countries, the risks of serious accidents and injury from a range of sources are increased, for example, by the use of toxic chemicals banned or restricted in other countries, unsafe techniques for chemical application or equipment use, the absence or poor maintenance of equipment, and lack of information available to the worker on the precautions necessary for minimizing risks during handling of agrichemicals, livestock, and machinery.

It is estimated that 70% of all child laborers (150 million) work in agriculture, which affects education, development, and long-term health. In addition to improving occupational health and safety, intersectoral action is needed to reduce and protect child labor through mechanisms such as access to education and health, poverty alleviation, and enforcement of child labor laws.

Emerging infectious diseases. Emerging and reemerging infectious diseases, including pandemic HIV/AIDS and malaria, are among the leading causes of morbidity and mortality worldwide [Global Chapters 1, 3, 5, 6, 8; SSA Chapter 3]. The incidence and geographic range of these infectious diseases are influenced by the intensification of crop and livestock systems, economic factors (e.g., expansion of international trade and lower prices), social factors (changing diets and lifestyles), demographic factors (e.g., population growth), environmental factors (e.g., land use change and global climate change), microbial mutations/evolution, and the speed with which people can travel around the globe. Serious socioeconomic impacts can arise when diseases spread widely within human or animal populations (such as H5N1), or when they spill over from animal reservoirs to human hosts; farming intensification often increases these risks. Even small-scale animal disease outbreaks can have major economic impacts in pastoral communities.

Future Challenges and Options for Improving Human Health through AKST

Malnutrition. Adequate nutrition requires a range of interrelated factors to be in place including food security, access to adequate supplies of safe water, sanitation, and education. AKST should be seen as a primary intervention to improve nutrition and food security, through development and deployment of existing and new technologies for production, processing, preservation, and distribution of food [CWANA; ESAP; Global Chapters 2, 3, 5, 8; LAC; NAE; SSA]. For example, evidence is beginning to accumulate that breeding biofortified crops may help address some human micronutrient deficiency and improve amino acid composition in major staples; use of targeted fertilizers, such as zinc, selenium, and iodine, on soils low in these essential human nutrients to correct deficiencies. Developing environmentally sustainable, food-based solutions to under-nutrition should be a priority. In both local and national food systems, policies and programs to increase crop diversification and dietary diversity will help achieve food security.

Dietary-related chronic diseases. There are well established mechanisms and tools for monitoring community nutrition status. These need to be used systematically to improve surveillance systems for both under- and over-nutrition, and of chronic disease rates, to ensure that governments appropriately address the rapidly changing nature of nutrition-related diseases in each country. Strategies for tackling the rises in overweight, obesity, and non-communicable diseases are needed in all world regions. Policies that simply rely on public health education and changing individual behaviors have been ineffective. Tackling nutrition-related chronic disease requires coordinated, intersectoral policy responses that include public health, agriculture, and finance ministries, as well as food industry, consumer organizations, and other civil society participation [Global Chapter 3; NAE].

There are often tensions between agricultural food policy and population health improvement goals. Despite claims that consumers determine the market, the actual health needs of consumers are seldom the driving factors in production decisions and agricultural policies [Global

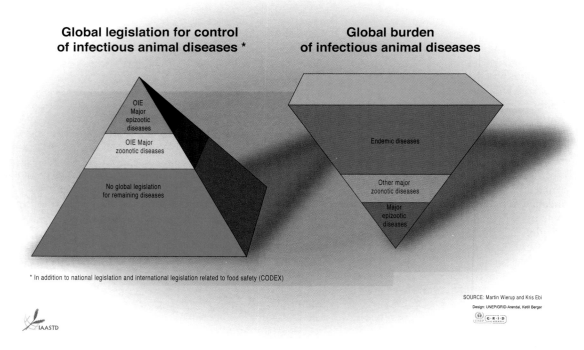

Figure SR-HH1. *Global legislation concerning, and global burden of, infectious animal diseases.*

Chapter 3; NAE]. Future AKST needs to refocus on consumer needs and well-being, for example the importance of diet quality and diversity should be main drivers of production and not merely quantity or price. Fiscal policies should take into account impacts on public health. Agricultural subsidies, sales taxes and food marketing incentives or regulations could be refocused to improve nutrition and public health as a primary aim, for example by promoting production and consumption of more healthy foods such as fruits and vegetables. AKST could improve dietary quality by regulating healthy product formulation through legislation or taxation (e.g., higher sales tax for food/foodstuffs known to cause adverse health effects, or limiting quantities of specific foods). Regulation may be necessary if voluntary industry codes are unsuccessful as has been the case in Sweden (banning of the use of transfats in processed foods) and the UK (reducing quantities of salt in processed foods). Other options for tackling nutrition-related chronic diseases include international agreements on and/or regulation of food labeling and health claims of products to ensure the marketing and labels are scientifically accurate and understandable for all consumers [Global Chapters 1, 3; NAE Chapter 2]. Such intersectoral polices should be designed and implemented alongside local and national public health action to maximize impact.

Food safety. AKST, along with strengthening and improving public health and environmental systems, can help ensure animal health, plant health, and food safety [CWANA Chapter 5; ESAP Chapter 3; Global Chapters 2, 5, 6, 7, 8; LAC Chapters 2, 3; NAE Chapters 2, 4; SSA Chapter 2]. This requires concerted efforts along the food chain, taking

a broad agroecosystem health approach. Examples include good agricultural practices (GAPs) and good manufacturing practices, integrated pest management, biological control of pests, and organic farming. These approaches, along with regulatory frameworks, can inform effective and safe pest and crop management strategies to manage the risks associated with pathogen contamination of foods. Implementing GAPs may help developing countries cope with globalization without compromising sustainable development objectives. Hazard analysis [risk assessment and food chain traceability] can enhance biosecurity and biosafety, disease monitoring and reporting, input safety [including agricultural and veterinary chemicals], control of potential foodborne pathogens, and traceability. Sanitation systems throughout the food production chain are integral to managing the risks associated with pathogens. Also needed is effective education of consumers in proper food handling and preparation.

However, AKST can increase the risks of food safety when technologies are applied without effective management of possible health risks. An example is the increasing use of treated wastewater in water-stressed agricultural systems in developing countries, where local communities have experienced increased rates of diarrheal diseases when either technologies or pathogen-contaminated wastewater outputs were used without effective controls.

Constraints to fuller deployment of current technologies and policies to improve food safety and public health include a wide and complex variety of factors (including market, trade, economic, institutional, and technical). There is a need to establish effective national regulatory standards and liability laws that are consistent with international best practices, along with the necessary infrastructure to ensure compliance, including sanitary and phytosanitary surveillance programs for animal and human health, laboratory analysis and research capabilities (such as skilled manpower and staff for research), and need-based and on-going training and auditing programs [CWANA Chapter 5; ESAP Chapter 3; Global Chapters 6, 7, 8; LAC Chapters 2, 3; NAE Chapters 2, 4; SSA Chapter 2].

Agrochemical exposure is of increasing concern [CWANA Chapter 5; ESAP Chapters 2, 3; 5; Global Chapters 3, 5, 6, 7, 8; LAC Chapters 1, 2, 3; NAE Chapters 2, 4; SSA Chapters 2, 3]. Use of agrochemicals is growing faster in developing than developed countries. Environmental and food safety impacts from agrochemicals, both positive and negative are determined by the conditions of use. Although there is no global mechanism to track pesticide-related illnesses, estimates of the number of possible cases and health costs are high, particularly in many developing countries without health insurance and universal health care.

Appropriate use of AKST can help prevent adverse health impacts along the food chain [CWANA Chapter 5; ESAP Chapter 3; Global Chapters 6, 7, 8; LAC Chapter 1; NAE Chapter 2; SSA Chapters 2, 3]. Place-based and participatory deployment of current (such as precision agriculture and bioremediation) and development of new technologies (such as biosensors) can reduce the risks associated with agrochemicals. Supply chain management presents a particular challenge in many less developed countries (LDCs), where the supply chain is characterized by limited coordination between farmers, traders, and consumers, poor infrastructure, and insufficient cold storage systems. Other challenges include harmonization of national and international regulations establishing upper levels of intake of nutrients and other substances, implementation of international treaties and recommendations, and improvement of food safety without creating barriers for poor producers and consumers. Implementation of these options requires major public and private research and development investments.

Occupational health. Agriculture is traditionally an underregulated sector in many countries and enforcement of any safety regulations is often difficult due to the dispersed nature of agricultural activity and lack of awareness of the extent of the hazards by those concerned. Few countries have any mechanism for compensation of occupational ill health. Current treaties and legislative frameworks, for example for agrichemicals, are not working. Improving occupational health in agriculture requires a greater emphasis on prevention and health protection, tackled through integrated multi-sectoral policies which must include effective national health and safety legislation (including child labor laws), and AKST which explicitly minimises health risks of agricultural workers. For example, health risks associated with pesticide use could be reduced through investment in pesticide reduction programs which could include incentives for alternative production methods (such as organic), investment in viable alternatives such as integrated pest management, and harm minimisation including withdrawal of generic compounds of high toxicity, and effective implementation of national and international regulations to stop cross-border dumping of hazardous and banned products [Global Chapters 1, 2,

3, 6, 7, 8; NAE Chapter 2]. AKST is essential to develop and deploy safer machinery and equipment, and improved knowledge transfer is required to improve use of existing and new technologies and techniques, including safe use of machinery, and livestock handling.

Occupational health will only be prioritized when the full extent of the problem becomes clear. This requires improved surveillance and notification systems on occupational accidents, injuries and diseases especially in LDCs. Agricultural and rural development policies should address the need for conducting occupational health risk assessments in the short term which make explicit the trade-offs between benefits to production, livelihoods, environmental and human health. These should include an assessment of all the external costs, including those on human health, as part of sustainable livelihood and poverty reduction programmes. Implementation of more agroecological approaches may result in synergies where reduction of input costs can also lead to improved livelihoods and harm minimization [Global Chapters 2, 3].

Emerging infectious diseases. Most of the factors that contribute to disease emergence will continue, if not intensify, in the 21st century, with pathogens that infect more than one host species more likely to emerge than single-host species [Global Chapters 5, 6, 7, 8]. The increase in disease emergence will affect both developed and developing countries. Integrating policies and programs across the food chain can help reduce the spread of infectious diseases. Examples include crop rotation, increasing crop diversity, and reducing the density, transport, and exchange of farm animals across large geographic distances. Focusing on interventions at one point along the food chain may not provide the most efficient and effective control of infectious diseases. For zoonotic diseases, this requires strengthening coordination between veterinary and public health infrastructure and training. Identification of and effective response to emerging infectious diseases requires enhancing epidemiologic and laboratory capacity, and providing training opportunities [CWANA Chapter 5; Global Chapters 5, 6, 7, 8; NAE Chapter 4; SSA Chapter 3]. Additional funding is needed to improve current activities and to build capacity in many regions of the world.

Detection, surveillance, and response programs are the primary methods for identifying and controlling emerging infectious diseases. Early detection, through surveillance at local, national, regional, and international levels, and rapid [and appropriate] intervention are needed [CWANA Chapter 5; Global Chapters 5, 6, 7, 8; NAE Chapter 4; SSA Chapter 3]. Effective public health systems and regulatory frameworks are needed to support these activities, as well as diagnostic tools, disease investigation laboratories and research centers, and safe and effective treatments and/or vaccines. Although AKST under development will advance control methods, there is limited capacity for implementation in many low income countries. For animal diseases, traceability, animal identification, and labeling (with associate educational initiatives) are needed. Recent advances in collection and availability of climate and ecosystems information can be used to develop forecasts of epidemics across spatial and temporal scales [Global Chapter 6]. Increasing understanding of the ecology of emerging infectious diseases can be integrated with environmental data to forecast where and when epidemics are likely to arise. Combined with effective response, these early warning systems can reduce morbidity and mortality in animals and humans. Additional research, improved coordination across actors at all scales, and better understanding of effective implementation processes are needed [CWANA Chapter 5; Global 5, 6, 7, 8; LAC Chapters 2, 3; NAE Chapter 4; SSA Chapter 3]. Information and communication technologies are creating opportunities for faster and more effective communication of disease threats and responses [Global Chapter 6]. Integrated vector and pest management are effective in controlling many infectious diseases, including environmental modification, such as filling and draining small water bodies, environmental manipulation, such as alternative wetting and drying of rice fields, and reducing contacts between vectors and humans, such as using cattle in some regions to divert malaria mosquitoes from people [Global Chapters 6, 7, 8; NAE Chapter 4]. Because the relationships between agriculture and infectious disease are not always straightforward, greater understanding is needed of the ecosystem and socioeconomic consequences of changes in agricultural systems and practices, and how these factors interact to alter the risk of emerging diseases.

Ways forward require human health to be seen by all actors as an explicit goal to be tackled by AKST. This requires integration and mainstreaming of public health throughout agricultural policies and systems.

Natural Resources Management

Writing Team: Lorna Michael Butler (USA), Roger Leakey (Australia), Jean Albergel (France), Elizabeth Robinson (UK)

Soil, water, plant and animal diversity, vegetation cover, renewable energy sources, climate, and ecosystem services are fundamental capital in support of life on earth [Global Chapter 1]. Natural resource systems, especially those of soil, water and biodiversity, are fundamental to the structure and function of agricultural systems and to social and environmental sustainability [Global Chapter 3]. The IAASTD report focuses primarily on the agronomic use of natural resources. Extractive processes such as logging, wild harvesting of non-timber forest products, captive fisheries [SSA SDM], while recognized as being important, are only addressed minimally here as they have been the focus of other global assessments.

In many parts of the world natural resources have been treated as though unlimited, and totally resilient to human exploitation. This perception has exacerbated the conflicting agricultural demands on natural capital, as have other exploitative commercial enterprises [ESAP Chapters 2, 4; Global Chapter 1]. Both have affected local cultures and had undesirable long-term impacts on the sustainability of resources [NAE Chapter 4]. The consequences include: land degradation (about 2,000 million ha of land worldwide) affecting 38% of the world's cropland; reduced water and nutrient availability (quality and access) [Global Chapter 1]. Agriculture already consumes 70% of all global freshwater withdrawn worldwide and has depleted soil nutrients, resulting in N, P and K deficiencies covering 59%, 85%, and 90% of harvested area respectively in the year 2000 coupled with a 1,136 million tonnes yr^{-1} loss of total global production [Global Chapter 3]. Additionally, salinization affects about 10% of the world's irrigated land, while the loss of biodiversity and its associated agroecological functions [estimated to provide economic benefits of US$1,542 billion per year (Global Chapter 9)] adversely affect productivity especially in environmentally sensitive lands in sub-Saharan Africa and Latin America [CWANA Chapter 2; Global Chapter 1, 6; LAC Chapter 1; SSA Chapter 5]. Increasing pollution also contributes to water quality problems affecting rivers and streams: about 70% in the USA [Global Chapter 8]. There have also been negative impacts of pesticide and fertilizer use on soil, air and water resources throughout the world. For example the amount of nitrogen used per unit of crop output increased greatly between 1961 and 1996.

The severity of these consequences varies with geographic location and access to the various capitals. This complex of interacting factors often leads to reduced livelihoods and diminishing crop yields, and the further refueling of natural resource degradation, especially in marginal areas [CWANA Chapter 1; ESAP Chapter 4; Global Chapters 3, 6; SSA Chapter 5]. The degradation of natural resources is both biophysically and socially complex. Interrelated factors drive degradation, for example: commerce, population growth, land fragmentation, inappropriate policy, customary practices and beliefs, poverty and weak institutions (customary and property rights, credit for the poor, crop and livestock insurance), can all be drivers of degradation [SSA Chapter 5]. On the other hand, there are examples where agricultural practices have been developed to protect agroecosystems [LAC Chapter 1; SSA Chapter 5], while producing marketable commodities [Global Chapter 3]. Examples include terracing, watershed and habitat management, protection of vulnerable landscapes, pastoral systems [SSA Chapter 5], and micro-irrigation technologies [Global Chapter 3], and, more recently, policies promoting biocontrol, organic food production, and fair trade [CWANA Chapter 2; LAC Chapter 1]. Additionally, loss of genetic resources has been partially addressed by establishment of gene banks and germplasm collections [Global Chapter 3]. However, the overexploitation paradigm still dominates.

Challenges

To improve the productivity of agriculture and enhance sustainable rural development there is the need to:

1. Assess the trends in the loss of natural capital (soil, water, plant and animal diversity, vegetation cover, energy, climate, ecosystem services) due to over-exploitation.

2. Understand the factors resulting in lower environmental resilience and the failure to achieve optimum agricultural output by the rural poor.

3. Mitigate and reverse the severe impacts on the environment and the livelihoods of poor people, for example resolving loss of soil fertility, erosion, soil salinization, decreased water quality and availability, decreased biodiversity and ecosystem services.

4. Resolve the biophysically and socially complex issues of NRM using formal, local and traditional knowledge, and collective, participatory and anticipatory decision making with diverse stakeholders across multiple scales.

5. Adopt a holistic or systems-oriented approach, to capture the needs for sustainable production and to address

Proportion of water withdrawal for agriculture, 2001

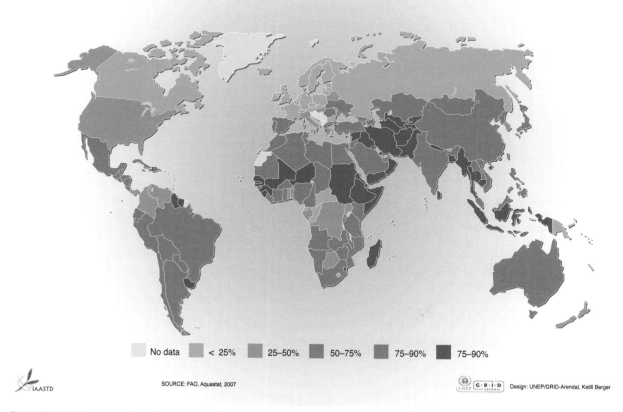

| No data | < 25% | 25–50% | 50–75% | 75–90% | 75–90% |

SOURCE: FAO, Aquastat, 2007

IAASTD

Design: UNEP/GRID-Arendal, Ketill Berger

Figure SR-NRM1. *Agricultural water withdrawals as proportion of total water withdrawals.*

the complexity of food and other production systems in different ecologies, locations and cultures so integrating food and nutritional security with natural resource management.

6. Determine who pays for the remediation of overexploitation and/or pollution of the natural resource system on which everyone depends.

Options for action relative to development and sustainability goals

The AKST available to resolve NRM exploitation issues like the mitigation of soil fertility depletion through synthetic inputs and natural processes, and the impacts of tillage on compaction and organic matter decomposition are often available and well understood. However, there is a need for greater knowledge and understanding of interactions between the agricultural system and the natural environment. Nevertheless, the resolution of natural resource challenges will demand new and creative approaches by stakeholders with diverse backgrounds, skills and priorities. Capabilities for working together at multiple scales and across different social and physical environments are not well developed. For example, farmer groups and civil society members have

rarely been involved in agricultural research, in shaping natural resource management policy, or in working partnerships with the private sector to achieve integrated natural resource management.

Causes of natural resource degradation and of declining productivity are multiple and complex. New AKST based on multidisciplinary approaches (biophysical, behavioral and social) is necessary for a better understanding of this complexity in NRM [NAE SDM Key Message 5; SSA Chapter 5].

Identify and resolve underlying causes of declining productivity embedded in natural resource mismanagement through the adaptation of existing technologies and the creation of innovative solutions.

*   *Land degradation and nutrient depletion*: The degradation of land is most often attributed to factors such as the loss of vegetation due to deforestation, overgrazing, land clearance, land abandonment, and inappropriate agricultural practices. It arises from population pressure, lack of appropriate technical support and knowledge, unavailability of inputs (fertilizers, water), conflicting social pressures, commercial incentives, sub-

Changes in available water

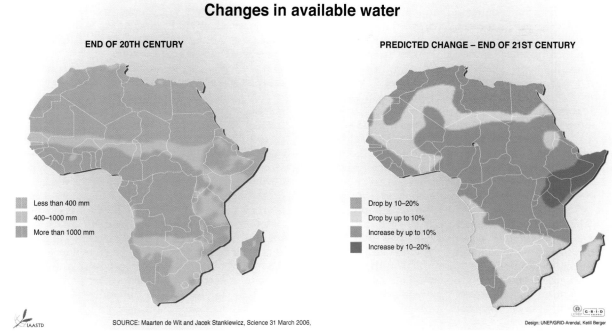

Figure SR-NRM2. *Changes in available water in Africa: end of 20th and 21st centuries.*

sidies and tariffs promoting non-sustainable practices, etc. Some proven technologies for mitigating land degradation include improved land husbandry, use of artificial and natural fertilizers, diversification and rotation of cropping systems, minimum or no-tillage, contour hedges, plowing, terracing and agroforestry practices, organic and conservation farming [CWANA Chapter 2; ESAP Chapter 5; Global Chapter 3; LAC Chapter 1; SSA Chapter 5].

- *Salinity and acidification*: Causes of salinity usually result from excessive irrigation and evaporation of soil moisture that draws up certain soil minerals, especially salt [CWANA Chapter 2]. Causes of acidification are related to overextraction of basic nutrient elements through continuous harvesting and inappropriate ferti-

lizer applications. The salinity problem can be reduced by minimizing irrigation application, and lowering water tables by appropriate tree planting, drainage systems; while acidification can be reduced by liming and addition of organic residues [Global Chapter 3; LAC Chapter 4].

- *Loss of biodiversity (above and below ground) and associated agroecological functions*: Loss of biological diversity results from repeated use of monoculture practices; excessive use of agrichemicals; agricultural expansion in to fragile environments; excessive land clearance that eliminates patches of natural vegetation; and neglect of indigenous knowledge and local priorities. This may be resolved by diversified farming systems; land-use mosaics; mixed cropping systems that integrate peren-

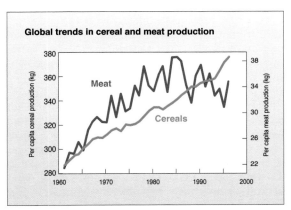

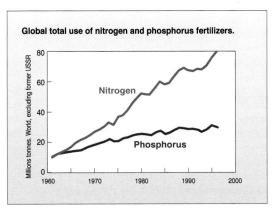

Figure SR-NRM3. *Annual global cereal production/annual global application of N (Source: Tilman et al., 2002).*

nials (cash crops or domestically important indigenous species); conservation farming and organic agriculture; integrated pest management; conserving or introducing biological corridors; controlling stocking densities; and ensuring pollination, seed dispersal, life cycles and food chains [Global Chapter 3; SSA Chapter 5].

- *Reduced water availability, quality and access*: Diffuse pollution from agriculture is a major factor in damaging water quality. Reduced water availability arises from river capture, exploitation of aquifers and ground water, drainage of wetlands, and deforestation. This can be countered by using appropriately constructed holding ponds, use of water-saving irrigation techniques, rainwater capture, riparian strips and erosion control, minimized use of agrichemicals, and improved efficiency in the use of manures and fertilizers [CWANA Chapter 2; Global Chapter 3; NAE Chapter 6].

- *Increasing pollution (air, water, land)*: This may be brought about by waste dumping, chemical accidents, unsuitable cultivation and land use practices that emit greenhouse gases, emissions from unregulated industry, etc. Pollution may be reduced by regulation (local, national, global); promotion of best practices for land/water use, e.g., carbon sequestration [CWANA Chapter 2; SR Part II: Climate Change]; reducing pesticide use; biological control; use of clean energy alternatives (biofuels, solar/wind power); etc. [Global Chapter 3; SR Part II: Bioenergy]

Strengthen human resources in the support of natural capital through increased investment (research, training and education, partnerships, policy) in promoting the awareness of the societal costs of degradation and value of ecosystem services.

- *Investment to promote awareness of resource resilience, protection and renewal*: This begins with creating understanding and awareness about sustainability issues and their impacts on various populations, environments and economies among national and international policy makers, donors, corporate business leaders and development agencies. This also requires public understanding of the issues. There are some good examples of two types of organizations that have brought part of the message to public attention. One is small organizations like Fair Trade and WWF; the other is global level policy, as exemplified by the Millennium Development Goals and the Kyoto Protocol to mitigate climate change. The latter have benefited from wide media attention. Agricultural sustainability would benefit similarly from media coverage conferring increased public understanding and support.

- *Investment in dissemination and implementation of promising multi-scale and commercially viable "packages" involving partnerships, technologies, appropriate practices, research and training programs.* Examples include Daimler-Chrysler's (Brazil) production of raw materials such as gums, oils, resins, and fibers for car manufacture by rural communities [Global Chapter 3]; ecoagriculture and ecotourism in which local communities, often with private sector partners, benefit from external interest in for example, local wildlife, unique

habitats, waterways, and forests; and use and protection of traditional knowledge and farmers' rights for better access to traditional foods, which can also enhance community empowerment [LAC Chapter 1].

- *Investment in research targeting natural resource resilience and renewal* and, simultaneously, strengthening local capabilities and ownership for wide scale adoption. Examples include rebuilding natural capital (replanting watersheds, soil fertility replenishment, replanting trees in the landscape); protection of water ways with riparian buffer strips; domestication of new tree crops through community action; wetland and swamp conservation; restoration of hydrological processes; and documenting and using traditional knowledge of natural resource conservation [ESAP Chapters 3, 4; Global Chapters 3, 6; LAC Chapters 1, 4; NAE Chapter 6].

- *Investment in research targeting mitigation of climate change and loss of biodiversity* [NAE Chapter 6]. Examples include developing better understandings of the role of biodiversity in agroecosystem functions and wildlife conservation through diversified farming systems that support local livelihoods [Global Chapter 3; SR Part II: Climate Change].

- *Investment in national, regional and global structures and partnerships* to protect natural resource data collections. Examples of secure data banks and collections include GEMS, IPGRI, and indigenous knowledge collections [see section on traditional knowledge and innovation; CWANA SDM; NAE Chapter 6].

- *Investment to promote improved models of extension and outreach* by engaging local people with scientists in participatory learning processes for NRM, and in adapting improved NRM technologies to local circumstances for a better informed public with the capabilities to diagnose, manage, and monitor natural resource issues and changes [LAC Chapter 5; NAE SDM; SSA Chapter 5].

- *Investment in cost-effective monitoring of the state of natural resources* to generate long-term trends and knowledge about the state of natural capital.

Promote agricultural production based on less exploitative NRM and strategies for resource resilience, protection and renewal through innovative processes, programs, policies and institutions.

- *Promote research "centers of AKST-NRM excellence"*. These would facilitate less exploitative NRM and strategies for resource resilience, protection and renewal through innovative two-way learning processes in research and development, monitoring and policy formulation [CWANA Chapter 2; NAE Chapter 6].

- *Develop a more multifunctional approach to agriculture* [NAE Chapter 6]. This can be achieved through integrating production of food crops within integrated farming systems that maintain environmental services such as carbon sequestration, soil organic management, water and nutrient cycling [NAE SDM]. This would benefit from the integration of local insights on land tenure and management regimes, gender-related patterns of resource access and control and participatory decision-making and implementation [ESAP Chapter 4;

Global Chapters 3, 5]. An example from West Africa demonstrates the possibility of improving the livelihoods of smallholder farmers by integrating trees into farming systems [Global Chapter 3], and the participatory domestication of traditionally important species [Global Chapter 3]. This example includes rural employment diversification (e.g., value adding) through postharvest activities [SSA Chapter 5].

- *Promote policy reform to instigate long-term improvements on existing agricultural land.* This will strengthen ecosystem services, prevent migration to forest and/or marginal lands, and agricultural land abandonment [Global Chapter 3; LAC Chapter 5].
- *Improve or establish land tenure institutions and policies.* This would include the promotion of common pool resource management and use (water, land, fisheries, forests); prevention of loss (or lack of clarity) of land rights and security, tenure inequity and lack of rights, particularly on the part of women and landless people [Global Chapter 3, 7; LAC Chapter 5; NAE SDM; SSA Chapter 5]; and appropriate natural resource allocation mechanisms, for example pricing, regulation, negotiation, enforcement, etc. Long-term improvements on existing agricultural land in order to prevent migration to forest and/or marginal lands, and agricultural land abandonment [Global Chapter 3].
- *The issue of who pays for environmental degradation is increasingly resolved by the principle "the polluter pays."* This is becoming an increasingly contentious issue as the population of the world grows more reliant on natural resources that are global public goods. Market mechanisms that address this challenge include Payment for Environmental Services (PES) that directly rewards improved management practices through transfers to those who protect ecosystem services from those who benefit. The Clean Development Mechanism links poor and rich countries through carbon trading. However, the costs of engaging in these mechanisms, and other market-based opportunities such as certification, are often beyond the reach of the poorest farmers [CWANA Chapter 2; Global Chapter 3; SSA Chapter 5].

Create an enabling environment that builds NRM capacity for concerted action among stakeholders and their organizations.

NRM stakeholders are likely to be more effective in shaping NRM policy when they have improved understanding of NRM issues, know the policy formulation process and have experience of working in partnership with public and private sectors [NAE SDM]. Multi-disciplinary teams have proven effective [CWANA Chapter 2; ESAP Chapter 4; LAC Chapter 4].

- *For marginalized groups* (e.g., women, youth, refugees, landless peoples, HIV-AIDS affected communities): Develop experiential learning, extension programs and primary and secondary education targeting improved NRM [Global Chapter 3; NAE SDM]. Important topics include use of information technology (IT) for NRM knowledge access, resource restoration, water-harvesting practices, land conservation and environmentally friendly farming technologies, collaborative

management, crop and animal domestication tools and strategies, low-input integrated approaches to farming (INRM, IPM), postharvest value-addition and marketing for business development, financial management, entrepreneurship and employment generation [ESAP Chapter 3; Global Chapters 3, 5; LAC Chapter 5; NAE SDM].
- *For community leaders and local government officials*: Develop capabilities that build capacity for multi-stakeholder partnerships [NAE Chapter 6], NRM leadership skills [Global Chapter 3] including IT capabilities. Important topics include land tenure policy; conflict resolution, feasibility planning, impact assessment, participatory group processes for natural resource management, restoration and recycling; financial management, entrepreneurship and employment generation; NRM strategies and technologies [Global Chapters 3, 5; LAC Chapter 5; NAE SDM].
- *For national and international policy makers*: Initiate learning opportunities to better understand the importance of IT connectivity and skill development, local and traditional knowledge in all aspects of NRM for agricultural research and development [Global Chapters 3, 5; SR Part II: TKI]. Additionally, promote models of extension and outreach that engage local people in participatory learning processes for NRM, and in adapting improved NRM technologies to local circumstances and needs, e.g., farmer organizations, farmer-to-farmer extension, participatory plant breeding [Global Chapter 3].

Facilitate natural resource management partnerships for different purposes to enhance benefits from natural resource assets for the collective good and to mitigate against natural hazards.

NRM partnerships are beneficial for landscape management and planning, technology and market development, policy development, research and rural development. AKST can support innovative partnerships across institutions for multi-stakeholder NR management.

- *At local, national, regional and international levels, create local-global collaborative research and development partnerships, based on mutual understanding, trust and goals.* Appropriate partners may include public and private sector representatives. In commercially oriented partnerships, there should be recognition of the development of IP and other mechanisms that benefit local partners and communities [ESAP Chapters 3, 4; Global Chapter 3; LAC Chapter 4].
- *Create partnerships and networks involving NGOs, CSOs, farmer field schools, government, private sector to build on shared knowledge and decision-making.* This may include training and mentorship to optimize implementation and outcomes. Long-term partnerships are essential for ensuring enduring capacity to benefit the collective good [Global Chapter 3; LAC Chapter 4; NAE SDM].
- *Ensure that each partner's contributions, together, represent the total needs of the partnership.* Trained facilitators can help strengthen the capacity of multi-stakeholder partnerships.
- *Examine and implement policies that encourage constructive NRM partnerships.* This would include limit-

Table SR-NRM1. *Globalization and Localization Activities.*

Globalization	Localization
Tropical plantations for export markets	Traditional subsistence agriculture
International commodity research by CGIAR	National research by NARS
	National extension services
Green Revolution	NGOs and CBOs
Agribusiness for fertilizers/pesticides and seeds	Farmer training schools
	Participatory Rural Appraisal
Multinational companies for commodity trade	Participatory domestication and breeding
WTO trade agreements	Fair trade
Biopiracy	Water-user associations
Biotechnology	Promotion of indigenous species/germplasm
	Equity and gender initiatives
	Recognition of farmer/community
	IPR
	Agroforestry for soil fertility management

ing or removing policies that constraint these partnerships [LAC Chapter 4; NAE Chapter 6].

Connect globalization and localization pathways that link locally generated NRM knowledge and innovations to public and private AKST to achieve more equitable and sustainable rural development.
Since the mid-20th century, globalization has been a dominant force in formal AKST. Public sector agriculture research, international trade and marketing, and international policy have been influential forces shaping globalization. Localization initiatives (Global Chapter 3; NAE Chapter 6) have come from the grassroots of civil society and involve locally based innovations that meet local needs of people and communities. Some current initiatives are drawing the two pathways together in ways that promote local-global partnerships for expanded economic opportunities. This is particularly true in the developing world in relation to the sustainable use of natural resources in agriculture [Global Chapter 3; NAE Chapter 6]. Natural resource management initiatives that illustrate how to bring localization and globalization together include:
- Promotion of customary foods to meet the needs and priorities of local people for self sufficiency, nutritional and food security, income generation and employment [Global Chapter 3].
- Domestication and commercialization of indigenous food-related plants and animal species [Global Chapter 3].

Global initiatives for sustainable development have brought attention to NRM issues at local and global levels, and have been effective in triggering the formation of civil society organizations, thereby stimulating new linkages with regional and/or global partners. Since the onset of the millennium some of these include: The Cartagena Protocol on Biosafety to the Convention on Biological Diversity (Montreal, Canada) in 2001, the International Treaty on Plant Genetic Resources for Food and Agriculture (Rome, Italy) in 2001, the World Summit on Sustainable Development (Johannesburg, South Africa) in 2002, and the World Food Summit (Rome, Italy) in 2002. Similarly, several international and regional assessments of relevance to NRM have promoted sustainable practices and people-oriented policies for addressing these issues. Some of these include: Millennium Ecosystem Assessment (2005); Intergovernmental Panel on Climate Change (1990, 1992, 1994, 2001, 2006); Comprehensive Assessment of Water in Agriculture (2007); Global Environmental Outlook; European Union Water Initiative; and European Union Soil Initiative.

Ways forward
Natural resource management is central to agricultural production and productivity, maintenance of critical ecosystem services and sustainable rural livelihoods. Agriculture represents one important management option, which when carried out in harmony with the landscape, can be beneficial to a wide range of stakeholders at all levels of community development [NAE-SDM]. It is evident that the severity of uncontrolled exploitation of natural capital is having major negative impacts on the livelihoods of both rural and urban people. By drawing down so severely on natural capital, rather than living on the interest, we are jeopardizing future generations. The challenges can be resolved if AKST is used and developed creatively with active participation among various stakeholders across multiple scales. This must be done in order to reverse the misuse of natural capital and ensure the judicious use and renewal of water bodies, soils, biodiversity, ecosystems services, fossil fuels and atmospheric quality for future generations.

Trade and Markets

Writing Team: Dev Nathan (India), Erika Rosenthal (USA), Joan Kagwanja (Kenya)

The challenge of targeting market and trade policy to enhance the ability of agricultural and AKST systems to strengthen food security, maximize environmental sustainability, and support small-scale farmers to spur poverty reduction and drive development is immediate. Agriculture is a fundamental instrument for sustainable development; about 70% of the world's poor are rural and most are involved in farming. National policy needs to arrive at a balance between a higher prices which can benefit producers and lead to a more vibrant rural economy, and lower prices, which, although volatile on the international market, can improve food access for poor consumers. The steep secular decline in commodity prices and terms of trade for agriculture-based economies has had significant negative effects on the millions of small-scale resource-poor producers [ESAP Chapter 3; Global Chapter 7]. Structural overproduction in NAE countries has contributed to these depressed world commodity prices. This is also a challenge in many developing country markets where overproduction of tropical commodities, particularly through the emergence of new producers who are willing to accept lower returns than established producers, has led to price collapse.

Under these conditions, a "business as usual" trade and market policy approach will not advance IAASTD objectives. There is growing concern that developing countries have opened their agricultural sectors to international competition too extensively and too quickly, before basic institutions and infrastructure are in place, thus weakening agricultural sectors with long-term negative effects for poverty, food security and the environment. Reciprocity of access to markets (sometimes referred to as a "level playing field") between countries at vastly different stages of agricultural development does not translate into equal opportunity [ESAP Chapter 3].

Agricultural trade offers opportunities for developing countries to benefit from larger scale production for global markets, acquire some commodities cheaper than would be possible through domestic production, and gain access to new forms of AKST and equipment (e.g., fertilizers, high-yield seed varieties, pump sets, etc.) not produced domestically. Agricultural trade, thus, can offer opportunities for the poor, but there are major distributional impacts among countries and within countries that in many cases have not been favorable for small-scale farmers and rural livelihoods. The poorest developing countries are net losers under most trade liberalization scenarios.

Trade policy reform aimed at providing a fairer global trading system can make a positive contribution to the alleviation of poverty and hunger. Approaches that are tailored to distinct national circumstances and different stages of development and target increasing the profitability of small-scale farmers are effective for reducing poverty in developing countries [CWANA; ESAP; Global; LAC; SSA].

Flexibility and differentiation in trade policy frameworks (i.e., "special and differential treatment") will enhance developing countries' ability to benefit from agricultural trade; pursue food security, poverty reduction and development goals; and minimize potential dislocations associated with trade liberalization. The principle of non-reciprocal access, i.e., that the developed countries and wealthier developing countries should grant non-reciprocal access to countries less developed than themselves, has a significant history and role to play in trade relations to foster development. Preferential market access for poorer developing countries, least developed countries and small island economies will be important.

Global Challenges

For many developing countries sustainable food security depends on local food production, while for some arid and semiarid countries with limited natural resources bases increased food security will require increased trade. Ensuring policy space for all these countries to maintain prices for crops that are important to food security and rural livelihoods is essential. Agricultural policies in industrialized countries, including export subsidies, have reduced commodity prices and thus food import costs; however this has undermined the development of the agricultural sector in developing countries, and thus agriculture's significant potential growth multiplier for the whole economy. Reducing industrialized countries' agricultural subsidies and other trade distorting policies is a priority, particularly for commodities such as sugar, groundnuts and cotton where developing countries compete. Commitments to reducing dumping, or the sale of commodities at below the cost of production thus undermining national food production and marketing channels are equally important.

Agricultural trade is increasingly organized in global chains, dominated by a few large transnational buyers (trading companies, agrifood processors and companies involved in production of commodities). In these globalized chains primary producers often capture only a fraction of the international price of a trade commodity, so the poverty reduction and rural development effects of integration in

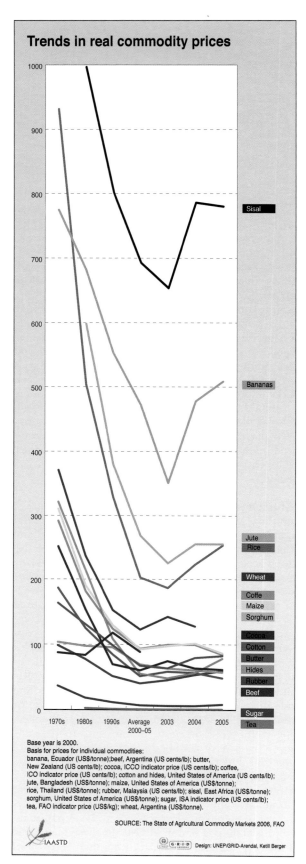

Figure SR-TM1. *Trends in real commodity prices.*

"*What are you complaining about? It's a level playing field.*"

Figure SR-TM2. *Level playing field.*

global supply chains have been far less than optimal [ESAP; NAE; Global]. Building countervailing negotiating power, such as farmer cooperatives and networks, will be important to help resource poor farmers increase their share of value captured.

Agriculture generates large environmental externalities including accelerated loss of biodiversity and ecosystem services such as water cycling and quality, increased energy costs and greenhouse gas emissions, and environmental health impacts of synthetic pesticides [ESAP Chapter 3; Global; NAE]. Many of these impacts derive from the failure of markets to value and internalize environmental and social harms in the price of traded agricultural and other products, or to provide incentives for sustainability. AKST has great potential to reverse this trend, aiding in the improvement of natural resource management and the provisioning of agroenvironmental services.

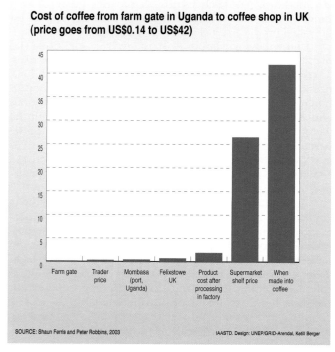

Figure SR-TM3. *Cost of coffee from farm gate to coffee shop.*

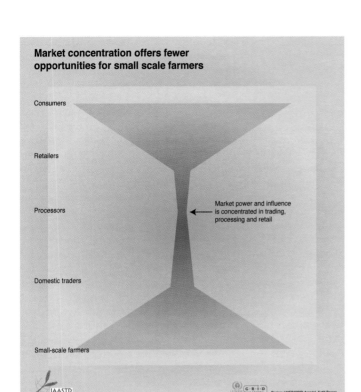

Market concentration offers fewer opportunities for small scale farmers

Consumers

Retailers

Processors

Market power and influence is concentrated in trading, processing and retail

Domestic traders

Small-scale farmers

IAASTD

UNEP/GRID-Arendal Design: UNEP/GRID-Arendal, Ketil Berger

Figure SR-TM4. *Market concentration offers fewer opportunities for small-scale farmers.*

Finally, improved local, national and global governance will enhance the ability of AKST systems to maximize agriculture as a driver for development. Governance is weakest in many agriculture-based developing countries, and governance of the agricultural sector is weak compared to other sectors. Enhanced global governance is also needed to support national sustainable development agendas.

Synthesis of priority challenges across regions
Many of the urgent challenges reported in the IAASTD are widely shared across the developing regions, or indeed, as in the case of climate and water crises, around the world. *Food security* is a priority agricultural trade policy challenge across the developing South. Trade policies designed to ensure sufficient levels of domestic production of food (not just sufficient currency reserves to import food) are an important component of food security and sovereignty strategies for many countries [CWANA; ESAP; LAC]. Approaches to balance domestic production with food stocks and foreign exchange reserves are noted in ESAP. A number of regions express significant concern over whether smaller economies would have sufficient foreign exchange reserves to cover increased food imports in light of declining terms of trade, and volatile international prices to import food [ESAP; SSA].

Additionally developing countries face significant new regulatory costs related to international trade. Tariff revenue losses have not been made up by other, domestic tax collections; tariffs used to represent a significant percentage of tax revenues in many developing countries. There are concerns that the high costs of regulatory measures to comply with

sanitary and phytosanitary standards will divert resources form national food and animal safety priorities. Investments to implement these standards should be approached as part of improvements needed to protect local populations from food-borne diseases and not only to comply with trade regulations.

Increased technical and financial assistance, as contemplated in the SPS Agreement, will be required to build and improve developing countries' own systems of quality control for meeting health and safety standards. Small producers, in particular, need technical, financial and management support to improve their production to meet health and safety standards.

Improving small scale farmers' linkages with local, urban and regional markets, as well as international markets, is noted across the developing country regions. *Enhancing regional market integration* to increase the size of markets (creating more constant demand and less price volatility), and negotiate from common platforms is a priority in SSA, LAC and ESAP. Assisting the small-scale farmer sector to access markets on more favorable terms, and capture greater value in global chains is emphasized [CWANA; ESAP; LAC; SSA].

Promoting investment for local value addition to increase diversity and competitiveness of agricultural products and generate off farm rural employment is a priority across the developing regions. It is widely noted that tariff escalation in industrialized countries has made it more difficult to stimulate investment in local value addition, exacerbating terms of trade problems [ESAP; LAC; SSA]. Concerns over preference erosion are also widespread [CWANA; Global; LAC; SSA].

The expansion of the agricultural landscape into forested areas and the potential for land planted for biofuels feedstocks to displace food crops and increase deforestation is a concern across the regions. Concerns about the vulnerability of agriculture to climate and water crises, equitable risk management and adaptation approaches, and the urgency of focusing AKST to reduce the environmental footprint of agriculture, emerge as clear global priorities [CWANA; ESAP; Global; LAC; NAE; SSA].

There is a concern expressed in many regions that intellectual property (IP) regimes have contributed to a shift in AKST research and development away from public goods provisioning. IP rights may restrict access to research, technologies, and genetic materials, with consequences for food security and development [ESAP; Global; LAC]. Improving the equitable capture of benefits from AKST systems is a priority in LAC and other regions. There often is a trade-off between rewarding the development of AKST through IP rights and, inhibiting dissemination and utilization. Countries may consider regional and bilateral cooperation in the formulation of national IPR systems and removing IPR from the ambit of WTO trade rules. Allowing greater scope to more effectively addressing situations involving traditional knowledge and genetic resources in international IP regimes would help advance development and sustainability goals.

Finally, the need to significantly improve the domestic policies for sustainable agricultural development to advance IAASTD objectives is noted across the developing South [CWANA; ESAP; Global; LAC; SSA]. This includes increasing the security of access and tenure to land and resources; targeting AKST research, development and delivery to meet the needs of small-scale farmers; and increasing investments in infrastructure such as post-harvest capacity, market feeder roads, and information services. Collective and individual legal rights to land and productive resources, especially for women, indigenous people and minorities, are emphasized in order for these groups to benefit from opportunities created by agricultural trade.

Options for Action to Advance Development and Sustainability Goals

This section discusses approaches to maximize the ability of trade and market policy options to facilitate targeted AKST to increase the agricultural sector's ability to deliver multiple public goods functions. There are important synergies and tradeoffs between policy options that merit special consideration. Potential liberalization of biofuels trade is a clear example, presenting tradeoffs between food security, greenhouse gas (GHG) emission reductions, and rural livelihoods which need to be carefully assessed for different technologies and regions, and is addressed at the end of this section [SR Part II: Bioenergy].

International trade policy options

Trade policy approaches to benefit developing countries include, among other measures, the removal of barriers for products in which they have a comparative advantage; reduced tariffs for processed commodities; deeper preferential access to markets for least developed countries, and targeted AKST research, development and dissemination for the small farm sector to advance development and sustainability goals.

Policy flexibility to allow developing countries to designate "special products," crucial for food security, livelihood and development needs as special products for which agreed tariff reductions will not be fully applied, are critically important to advance development and sustainability goals. This gives developing countries an important tool to protect these commodities from intensified import competition, until enhanced AKST, infrastructure and institutional capacity can make the sector internationally competitive. Similarly the special safeguard mechanism [SSM], designed to counter depressed prices resulting from import surges, is an important trade policy tool to avoid possible damage to domestic productive capacity. At the household level depressed prices can mean inability to purchase AKST, the need to sell productive assets or missed school fees [ESAP; Global]. World Trade Organization country categories that better reflect the heterogeneity of developing countries' food security situations could help ensure that no food insecure country is denied use of these mechanisms.

The elimination or the substantial reduction of subsidies and protectionism in industrialized countries, especially for commodities in which developing countries compete such as sugar, groundnuts and cotton is important for small-scale farm sectors around the world. Similarly, plurilateral commitments from major exporting countries to ensure that there is no trade at prices below the full cost of production have been put forward as an option to discipline dumping (which can cause significant damage to small-scale produc-

ers). There is need for increased attempts to find alternate uses for these commodities, e.g., fruit coating with *lac*, or bio-fuel from palm oil. International commodity agreements and supply management for tropical commodities, with improved governance mechanisms to avoid problems of free-riding and quota abuse are receiving renewed consideration to address price-depressing structural oversupply. International trade and domestic policies need to manage orderly shifts in production centers, enabling producers in high-cost centers to shift, without the destitution that can be brought about by pure market-induced transitions. Elimination of escalating tariffs in industrialized countries would help encourage value-added agroprocessing to help create off-farm rural jobs and boost rural livelihoods. It would also assist in diversifying fisheries production and exports toward value-added processing, reducing fishing pressure on dwindling stocks.

Increasing support for public sector research to deliver public goods AKST outputs is important to meet development and sustainability goals, along with implementation of farmers' rights to seeds to enhance conservation of agricultural biodiversity and associated informal AKST. Administering effective mechanisms to protect traditional and local knowledge remains a challenge [ESAP Chapter 3; Global; LAC.

Replacing revenues lost as a result of reduced import tariffs is essential to advance development agendas. If countries are not able to make up the revenue difference with other taxes (i.e., consumption taxes that are economically more efficient but can be administratively and politically difficult to collect) the pace of tariff reduction could be reconsidered. Increased Aid for Trade and development assistance commitments will also be necessary. Priorities should be determined on an individual country basis, including AKST targeted to improve competitiveness; strengthen institutional capacity for trade policy analysis and negotiation; and cover costs of adjustment for measures that have already been implemented. (Industrialized countries have a right and an obligation to compensate their own losers as well.)

National trade and market policy issues

National agricultural trade policy to advance sustainability and development goals will depend upon the competitiveness and composition of the sector. Advice to developing countries has tended to focus on promoting opportunities

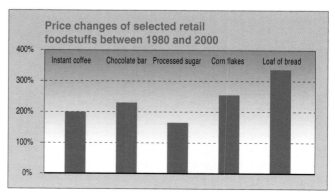

Figure SR-TM5a. *Price change of selected retail foodstuffs*

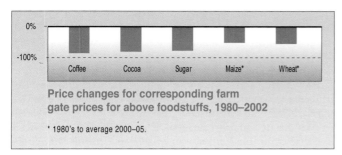

Figure SR-TM5b. *Percentage of retail value paid to primary producer.*

for increased exports to international markets (traditional and non-traditional crops) rather than enhancing competitiveness of import substitutes or market opportunities in domestic and regional markets; greater balance among these policy approaches may be indicated.

It is increasingly recognized that developing countries at an earlier stage of agricultural development may require some level of import protection for their producers while investments are made to improve competitiveness. State trading enterprises in developing countries (with improved governance mechanisms to reduce rent-seeking) may provide enhanced market access for marginalized small-scale farmers in developing countries, creating competition in concentrated export markets.

Developing countries benefit from *improved security of access and tenure to land and productive resources* (including regularization and expansion of land ownership by small-scale producers and landless workers), and increased research, development and effective delivery of AKST targeted to the needs of resource-poor producers. Strengthening social capital and political participation for the poor and vulnerable offer significant opportunities to reduce poverty and improve livelihoods. Legal rights and access to land and productive resources such as micro-credit and AKST, is key to improving equity and the ability of women, indigenous peoples and other excluded sectors to benefit from trade opportunities.

Options for accessing markets on more favorable terms

Better access to capital, local value addition and vertical diversification, improved infrastructure, AKST targeted to resource poor farmers, facilitation of farmer organization and collective action to take upscale-sensitive functions and alternative trading channels can help increase the bargaining position of small producers within global chains [ESAP; Global; LAC; SSA].

Expanding access to *microfinance* is an option to allow small-scale producers to access AKST inputs and technologies, and improve investment and asset building. This includes products and services offered by financial institutions as well as credit and other services offered by value chain actors. Newer financial services and products, such as crop or rain insurance, can help reduce risks associated with adopting new technologies, transitioning to agroenvironmental practices, and innovating production and marketing methods.

Supporting development of *fair trade and certified organic agriculture* offers an alternative set of trading standards to mainstream commodity markets that can improve the environmental and social performance of agriculture, and provide greater equity in international trade by providing favorable and stable returns to farmers and agricultural workers. Commitments to source fair trade products, and support for fair trade networks for basic foodstuffs and south-south sales, are promising approaches. Certified organic agriculture is value-added agriculture accessible to resource poor farmers who have extensive local production knowledge and capacity for innovation. Options to support the growth of organics include developing capacity in research institutions, crop insurance and preferential credit, and tax exemptions on inputs and sales. New business models and *private sector sustainable trading initiatives* apply these standards to mainstream trading operations via reducing the cost of certification and compliance for groups of small scale farmers; improving financial sustainability through buying relationship that better balance risk, responsibilities and benefits among the chain actors; and increasing information sharing and capacity building to increase business skills for producer organizations.

Market mechanisms to internalize negative and reward positive environmental externalities

Key trade and market policies to facilitate AKST's contribution to reducing agriculture's large environmental footprint include removing perverse input subsidies, taxing externalities, better definition and enforcement of property rights, and developing rewards and markets for agroenvironmental services.

Payments/reward for environmental services (PES) is an approach that values and rewards the benefits of ecosystem services provided by sustainable agricultural practices such as low-input/low-emission production, conservation tillage, watershed management, agroforestry practices and carbon sequestration. A key objective of PES schemes is to generate stable revenue flows that can help ensure long-term sustainability of the ecosystem that provides the service. To achieve livelihood benefits as well as environmental benefits, arrangements should be structured so that small-scale farmers and communities, not just large landowners, may benefit [Global; LAC; NAE].

Other policy approaches to address the environmental externalities of agriculture include *taxes on carbon and pesticide use* to provide incentives to reach internationally or nationally agreed use-reduction targets, tax exemptions for biocontrols to promote integrated pest management, and incentives for "multiple" functions use of agricultural land to broaden revenue options for land managers. [ESAP; Global; LAC; NAE] *Carbon-footprint labels* are an option to internalize the energy costs of agricultural production via the application of a market standard. Assistance to small-scale producers, especially tropical producers, to articulate their carbon rating will be key; in many cases, an integrated analysis of energy costs and GHG emissions from distant developing country production will be favorable [Global].

Identification and elimination of environmentally damaging subsidies, including fishery subsidies is a fundamental. *Fisheries subsidies* that fuel overexploitation and threaten the viability of many wild stocks and the livelihoods of fishing communities are an example. Options include investment in value-added processing, as well as subsidies for reduced fishing and for mitigating the negative social and economic consequences of restructuring the fisheries sector [Global Chapter 7].

Finally, improving interdisciplinary international cooperation on a wide range of agriculture and environmental issues is essential to advance development and sustainability goals. For example, a more comprehensive climate change agreement could include a modified Clean Development Mechanism to take fuller advantage of the opportunities offered by the agriculture and forestry sectors to mitigate climate change. The framework would include a comprehensive set of eligible agricultural mitigation activities, including: afforestation and reforestation; avoided deforestation using a national sectoral approach rather than a project approach to minimize issues of leakage; and a wide range of agricultural practices including zero/reduced-till, livestock and rice paddy management. Other approaches could include reduced agricultural subsidies that promote GHG emissions. Mechanisms that also encourage and support adaptation, particularly in regions that are most vulnerable such as in the tropics and sub-tropics, and that encourage sustainable development might also be included in a post-Kyoto climate regime [Global; NAE]. An efficient mechanism to handle interactions between multilateral environmental agreements and trade regimes is needed in order to ensure environmental and development concerns are not made secondary to trade rules.

Enhancing governance
Approaches to address the imbalance in trade relationships between small-scale producers and a limited number of powerful traders include the establishment of *international competition policy* such as multilateral rules on restrictive business practices, and an international review mechanism for proposed mergers and acquisitions among agribusiness companies that operate in multiple countries simultaneously. The creation of an independent agency to take up the mandate of the UN Center for Transnational Corporations could generate much needed information and analysis to support sustainable development agendas.

The quality and transparency of governance of AKST decision making is fundamental, including increased information and analysis for decision makers, and meaningful participation of all relevant stakeholders. *Strengthening developing country capacity* to analyze and identify options that are in their best interest, and play a full and effective role in the negotiation process, is a prerequisite for a positive and equitable outcome of trade negotiations. Increased Aid-for-Trade and other support will be necessary. Consideration may also be given to establishing national and regional teams of experts to analyze the interests of their stakeholder groups and recommend negotiating positions.

There is often *limited information* on the potential social, environmental and economic consequences to different sectors of society and regions of the world, of both proposed trade agreements and emerging technologies. Increased access to information requirements may be applied to the trade process, allowing for greater civil society access to information and participation in policy formulation [Global]. Analysis tailored to countries at different stages of development, and different characteristics of agriculture sectors and household economies can better inform policy choices to address development and sustainability goals. Developing better tools for assessing tradeoffs in proposed trade agreements includes increased use of strategic impact assessments (SIAs). SIAs aim to give negotiators and other interested stakeholders a fuller understanding of potential social, economic and environmental risks and benefits before commitments are made.

An intergovernmental framework for comparative technology assessments would increase information for decision makers on emerging technologies for agriculture, including, for example, nanotechnologies. This may include creation of independent international, regional or national bodies dedicated to assessing major new technologies and providing an early listening and warning system, or the establishment of a multilateral agreement to promote timely comparative technology assessment with respect to development and sustainability goals.

Traditional and Local Knowledge and Community-based Innovations

Writing team: Satinder Bajaj (India), Fabrice Dreyfus (France), Tirso Gonzales (Peru), Janice Jiggins (UK)

Traditional and local knowledge constitutes a vast realm of accumulated practical knowledge that decision makers cannot afford to overlook if development and sustainability goals are to be achieved [ESAP SDM; Global SDM; Global Chapter 3, 7; 8; NAE SDM; LAC Chapter 1]. Effective, sustainable technologies with wide scale application that have originated in local and traditional AKST are numerous and found worldwide. They include the use of Golden Weaver ants as a biocontrol in citrus and mango orchards (Bhutan, Viet Nam and recently with WARDA's assistance, introduced to West Africa); stone lines and planting pits for water harvesting and conservation of soil moisture (West African savannah belt); *qanats* and similar underground water storage and irrigation techniques (Iran, Afghanistan and other arid areas) [CWANA SDM]; tank irrigation (India, Sri Lanka); many aspects of agroforestry (3 million ha of rubber, cinnamon, damar agroforests in Indonesia) and current initiatives to domesticate indigenous tree species producing fruits, nuts, medicines and other household products [Global Chapter 3]. Many kinds of traditional and local AKST support wildlife and biodiversity and contribute to carbon and methane sequestration [Global Chapters 2, 3].

In numerous cases traditional and local AKST in collaboration with formal AKST and support services is empowering communities, maintaining traditional cultures and diets while improving local food sovereignty, incomes, nutrition and food security [Global Chapter 3]. Partly because the innumerable but diverse innovations resulting from local and traditional AKST are hard to present as statistical data they typically are overlooked, undervalued and excluded from the modeling that often guides AKST decision making [ESAP SDM; Global Chapters 2, 3].

Local and traditional agricultures work with genetic material that is evolving under random mutation, natural and farmer selection and community management [Global Chapter 2]. Even in unpromising soil and topographic conditions, as in the high Andes, local and traditional knowledge nurtured and managed germplasm that today is recognized as a center of origin of genetic diversity. Local and traditional strategies for *in situ* conservation can be highly effective in managing the viability and diversity of seed, roots, tubers and animal species over generations. [Global Chapter 3] The diversity gives local options and capacity for adaptive response that are essential for meeting the challenges of climate change [CWANA SDM; Global Chapters 2, 3].

Mobilizing these capacities in collaboration with formal science can generate AKST of more than local significance [Global Chapter 3]. Robust evidence indicates that it is the form of collaboration that determines the effectiveness of the resulting AKST in terms of development and sustainability goals [Global Chapters 2, 3, 4].

The nature of traditional and local knowledge

Traditional knowledge [Global Chapter 7]. The UN Convention on Biological Diversity refers to traditional knowledge, innovations and practices of indigenous and local communities embodying traditional lifestyles relevant for the conservation and sustainable use of biological diversity [Global Chapter 2]. More broadly, traditional knowledge is constituted in the interaction of the material and non-material worlds embedded in place-based cultures and identities [Figure SR TKI-1] [LAC SDM].

The local *Pacha* (mother earth) is a micro-cosmos, a representation of the cosmos at large. It is animated, sacred, consubstantial, immanent, diverse, variable, and harmonious. Within the local *Pacha* there is the *Ayllu* (Community in Quechuan and Aymaran languages). The *Ayllu* is comprised of three communities: people, nature, spirits. Throughout the agricultural calendar interaction within the *Ayllu* takes place through rituals and ceremonies. The place par excellence for the three communities to interact is the *chacra* (plot size: 1 to 2 ha). Harmony is not given, it has to be regularly procured through dialogue, reciprocity, redistribution and rejoicing flowing among the three communities. Nurturance and respect are fundamental principles in these exchanges. Knowledge created and transferred from another place by persons from outside the locality has to be instituted in the *chacra* through and in harmony with the dialogue among the members of the *Ayllu* and in conformity with the rituals and ceremonies that support such dialogue.

Local knowledge is a functional description of capabilities and activities that exist among rural actors in all parts of the world, including OECD countries [Global Chapter 2; LAC SDM]. Local stakeholders may engage in AKST activities typically (1) to compel acknowledgment of their knowledge and capacity for self-generated development by organizations and actors located elsewhere or (2) to reap benefits by fostering relations with non-local organizations and actors who need contextual, place-based knowledge in order to perform their own missions efficiently and profitably [Global Chapter 2]. Labels of geographical origin exemplify the first; the second is instanced by formal breeders and commercial organizations in the Netherlands who cooperate with Dutch potato hobby specialists in breeding and varietal selection; the farmers negotiate formal contracts which give them recognition and due reward for their intellectual contribution in all varieties brought to market [Global Chapter 2].

The dynamics of traditional and local knowledge. Traditional and local knowledge co-evolve with changes in their material and non-material environment. Any internal and external forces and drivers [including weather-related events] that threaten the loss of the material basis of traditional and local cultures and identities necessarily threaten traditional and local knowledge [CWANA SDM; ESAP SDM; Global Chapter 3].

Encounters between traditional and local knowledge actors and others

Encounters that support sustainability and development. There is a wealth of evidence of encounters between knowledge actors that have supported achievement of development and sustainability goals [ESAP Chapter 2; Global Chapters 2, 3, 4; LAC SDM; NAE Chapters 1, 4; NAE SDM].
- Participatory, collaborative methods and approaches have added value to the encounter between traditional/local knowledge actors and formal AKST actors. Farmer-researcher groups in the Andes for instance brought together members of CIP (an international research institute) for the development and testing of measures and varieties to control late blight in potatoes, not only increasing productivity but also addressing issues for instance of inter-generational equity and the sustainability of soil management. Collaboration among knowledge actors in the commercialization and domestication of tree [and other] wild and semiwild species in participatory plant breeding (PPB) and in value-added processing are creating new value chains selling into both niche and mass markets [Global Chapters 2, 3, 4]. Other examples include efforts made in a number of countries to invite traditional/local knowledge actors into rural schools (e.g., Thailand) and universities (e.g., Peru, Costa Rica) as teachers and field trainers; to incorporate local AKST in the curricula and experiments run by village-based adult education and vocational training centers (e.g., India); and to expand opportunities for experiment-based, farmer-centered learning [Global Chapter 2]. Modern ICTs show large potential for extending and augmenting these developments [Global Chapter 2].
- Encounters between traditional knowledge actors also can support sustainability and development [Global Chapter 3]. An example of fruitful encounters is given by the extension of rice cultivation in brackish water in coastal Guineas [Conakry and Bissau]. Migrants from the ethnic sussu met local ethnic balantes in Guinea Bissau around 1920 and, later on, local sussu (and also related ethnic baga) hired migrant balantes to implement

Figure SR-TK-1. *The Andean cosmovision.*

rice cultivation in Guinea Conakry where it is now regarded as a traditional knowledge [Global Chapter 2].

Encounters that threaten sustainability and development. Less favorable encounters have been associated mostly with AKST that focuses on objectives that are not shared by local people. Typically these have arisen in the context of the following circumstances:
- *Colonial disruptions* that continue in some parts of the world with lingering but strong influences. In some cases they serve to erode common property management regimes, leading to uncontrolled open access to natural resources and resource degradation [Global Chapter 4] or privatization of local people's land [Global Chapter 7].
- *Profit-seeking forces acting at the expense of multifunctionality.* Mechanisms to increase the accountability of powerful commercial actors to development and sustainability goals have been weak. In recent decades public information campaigns, shareholder activism and more effective documentation and communication of malpractices have begun to exert some pressure for change. Modern information and communication technologies have assisted these developments but the already poor and marginalized have less access to these means [Global Chapter 2].

- *Technical developments that assume rather than test the superiority of external knowledge and technologies* in actual conditions of use, conveyed by Transfer of Technology models of research-extension-farmer linkages [ESAP Chapter 2; Global Chapter 3, 7, 8]. Formal research agencies and universities have lagged behind in developing criteria and processes for research prioritization and evaluation that go beyond conventional performance indicators to include a broader range of criteria for equity, environmental and social sustainability developed by traditional people and local actors [LAC SDM]. Decision making processes in and the governance of formal institutes of science and research generally have excluded representatives or delegates of traditional peoples, poor local communities or women [LAC SDM] who only in exceptional circumstances have had a voice on governing boards, impact assessment panels, advisory councils and in technology foresight exercises. Their inclusion has required deliberate and sustained processes of methodological innovation, institutional change and capacity development [Global Chapter 2].

- *Misappropriation.* In some cases external actors have used without direct compensation the biological materials developed under local and traditional communities' management yet have largely ignored the knowledge and understanding that accompanied the *in situ* development of germplasm. The important public role of gene banks to return to local communities traditional germplasm that may have been lost at local levels has become more constrained under the evolution of Intellectual Property Rights regimes. Material transfer agreements in practice or law also may provide powerful public and commercial actors privileged access to this germplasm [Global Chapter 2].

- *Suppression of local knowledge, wisdom and identity.* In worst but far from rare cases educational curricula have been used deliberately to suppress traditional and local knowledge and identities. Inappropriate content or facilities in school-based education in some instances has worsened existing bias against attendance by traditional peoples or by girls and women [CWANA SDM; LAC SDM].

Asymmetries of power in institutional arrangements for AKST. The explanatory value of inequitable power relations has been demonstrated in the assessment of the positive and negative outcomes of encounters between knowledge actors in relation to development and sustainability goals. Formal AKST centers [CWANA SDM; ESAP SDM; LAC SDM], have privileged conventional systems of production; agroecological and traditional systems of production have been marginal in the R&D effort made [CWANA SDM; Global Chapter 3]. Knowledge actors based in formal research organizations have neglected development of accountability for the costs of some technologies—such as highly toxic herbicides and pesticides when applied in actual conditions of use [CWANA SDM; ESAP SDM] that have been borne disproportionately at local levels and often by the most marginalized peoples [Global Chapter 2; NAE].

A globalizing world. A globalizing world has offered opportunities that are welcomed and actively sought by tradition-

al and local people but also brought new risks, especially for the vulnerable and ill-prepared. Mutual misunderstanding across languages and other divides can undermine opportunities for collaboration especially when engagement is not mediated by inter-personal interactions but by impersonal bureaucracies, companies or commercial operations.

Persistent concerns for which as yet no lasting remedies have been found include the increasing competition for groundwater and river systems between local and non-local users [CWANA SDM—Farm structures and production systems], as well as the alienation of land and restriction of access to the habitats that have sustained and nurtured traditional and local communities' knowledge generation [ESAP SDM; Global Chapter 3]. While years of protest from indigenous peoples, community organizations and activist groups by the 1990s helped ensure that the principles of benefit sharing in the exploitation of local and traditional resources were written into international conventions such as the UN Convention on Biological Diversity, these lacked enforcement mechanisms. There has been a progressive restriction of communities' and farmers "rights to produce, exchange and sell seed". The freedom of states to recognize these rights is limited under UPOV 1991 and further limitations are proposed by some powerful commercial and government actors. The slow pace of adjustment of national varietal approval mechanisms for materials generated by farmers' organizations and through PPB has raised new challenges [Global Chapters 2, 3, 4].

Challenges

Institutionalization and affirmation of traditional and local knowledge [Global Chapter 7, 8]. Concerned actors in a number of countries have developed strategies at local to national levels to institutionalize and affirm traditional and local knowledge for the combined goals of sustainable agricultural modernization, NRM, social justice and the improvement of well-being and livelihoods [Global Chapter 3; LAC SDM, Chapter 5]. Robust examples include the *gram panchayat* [village councils] in India [ESAP SDM] and local water user associations [Global Chapter 3]. Currently some countries (e.g., Mali, Thailand) also are establishing policy frameworks that are congruent with the overall objectives of market-oriented sustainable development yet recognize the importance of traditional and local AKST capacities. The wider application or scaling up of such experiences faces strong and persistent challenges [Global Chapter 2].

Education. The more widespread application of collaborative approaches in AKST practices would require [a] complementary investments in the education of AKST technicians and professionals in order to strengthen their understanding of and capacity to work with local and indigenous individuals and communities; [b] support to curriculum developments that value and provide opportunity for field-based experience and apprenticeships under communities' educational guidance; [c] farmers' access to formal training to enable them to connect to innovations in agroecology [CWANA SDM; ESAP Chapter 4; Global Chapter 2; LAC SDM].

The valuation of traditional and local AKST [Global Chapter 7; NAE SDM, Chapter 1]. Certification and similar means of linking consumers and producers to traditional and local identities have been developed to give value in the marketplace to traditional and local knowledge and foods [ESAP SDM; Global Chapters 3, 4]. Some of the certified foods available today also include the "quality of life" values important to traditional producers or local communities [Global Chapter 3]. An increasing number of commercial actors in agrifood and agrochemical industries also are demonstrating their commitment to sustainable production and retailing through accreditation, auditing and traceability [Global Chapter 2, 3; LAC SDM].

Issues of laws, regulations and rights. It is recognized—yet not accepted at all policy levels—that innovations to secure rights for farmers, traditional people and citizens over germplasm, food, natural resources or territories are needed if combined sustainability and development goals are to be met [ESAP Chapter 3; Global Chapters 3, 7]. A number of countries (e.g., Mali), indigenous peoples (e.g., the Awajun, Peru) and local governments [e.g., various municipalities in the Philippines] have adopted the principles of food sovereignty as well as normative policy frameworks and regulations that differentiate their own needs and circumstances from the dominant global arrangements [Global Chapter 2; LAC SDM].

Options for action

Four key areas for action were identified:

(1) *Affirm local and traditional knowledge* [NAE SDM, Chapter 4] *by investment in the scientific, local and traditional conservation, developing and using local and traditional plants, animals and other useful biological materials, using advanced techniques as well as sophisticated application of participatory and collaborative approaches* [Global Chapter 8]. Specific investments include development of greater professional and organizational capacity at all levels for research and development with and for local and traditional people and their organizations [ESAP SDM; LAC SDM; NAE SDM] and support for multistakeholder AKST forums at all levels for building a shared understanding and collective vision among divergent interests [Global Chapter 7; LAC SDM; NAE SDM]. Options for affirmation include documentation and "archiving" of local and traditional people's knowledge products, knowledge generating processes and technologies—for instance in formal knowledge banks as well as in community-held catalogues of practices, designs and ancestral plant and animal genetic resources; and targeted support for in situ and ex situ conservation of crop, fish, forest and animal genetic resources [LAC SDM].

(2) *In education, give higher priority for agroecological and integrated approaches in primary through tertiary education and research* [Global Chapter 3, NAE SDM, Chapter 4]. Invest in a broader range of social sciences to understand and help design solutions to power asymmetries in AKST; arrange for effective encounters between knowledge actors and knowledge organizations [Global Chapter 2];

widen development of the role of local and traditional trainers in educational curricula and deepen and strengthen the educational options. Invest in occupational education and farmer-centered learning opportunities that are accessible and relevant for traditional and indigenous peoples and actively extend connectivity and ICTs to traditional and local knowledge actors [Global Chapter 3, NAE SDM, Chapter 4] and expand the coverage of the above.

(3) *Continue institutional innovation in systems such as Fair Trade, geographic identification and in value chains that shorten connections between producers and consumers* [ESAP Chapter 3; Global Chapter 3; NAE SDM]. Support the valuation of local and traditional knowledge. Develop culturally appropriate modes of assessing traditional and local AKST contributions to achievement of development and sustainability goals [Global Chapter 6]. Widen support of efforts to create local opportunity for domestication of wild and semiwild species [Global Chapter 3]. Support to conservation and evolution of local and traditional medicinal plants, knowledge of healing and health care systems [ESAP Chapter 3] as well as certification, regulation and marketing schemes that take account of traditional and local people's criteria and standards are options that make visible in the market places, societies and at policy levels the value of local and traditional knowledge.

(4) *Institutions, laws and regulations offer substantial options.*

- Decentralization and devolution of services; local government support to community-driven development [Global Chapter 7];
- Investment in research to underpin the design of methods and processes for integration of AKST decision-making at different scales [Global Chapter 8; NAE SDM, Chapters 3, 4];
- Follow-through on the Joint Indigenous People's Statements, 1999, 2007;
- Regional networking among community groups and traditional peoples' movements around pesticide and herbicide management [Global Chapter 2];
- Building co-responsibility for AKST outcomes and stronger, more effective mechanisms for enforcing these;
- Developing "best practice" procedures and processes for including traditional and local people in AKST research prioritization, technology assessments and evaluation [Global Chapter 3];
- Evolution of Intellectual Property concepts, rules, and mechanisms congruent with development objectives and the rights of local and traditional peoples. [ESAP SDM; Global Chapter 3, 7; NAE SDM];
- Institutional innovations at policy level in support of implementation of the CBD, UNECCO-Link;
- Access and Benefit-sharing Agreements [Global Chapter 3] and other systems for protecting Farmers' Rights [Global Chapter 7] and stronger coordination among such initiatives.

Women in Agriculture

Writing Team: Alia Gana (Tunisia), Thora Martina Herrmann (Germany), Sophia Huyer (Canada)

Gender, that is the socially constructed relations between men and women, is an organizing element of existing farming systems worldwide and a determining factor of ongoing processes of agricultural restructuring. Current trends in agricultural market liberalization and in the reorganization of farm work, as well as the rise of environmental and sustainability concerns are redefining the links between gender and development, as women not only continue to play a crucial role in farm household production systems, but also represent an increasing share of agricultural wage labor.

Since the first world conference on women (1975), the attention of decision makers has been attracted to the need for policies that better address gender issues as an integrative part of the development process. Although progress has been made in women's access to education and employment, we must recognize that the largest proportion of rural women worldwide continues to face deteriorating health and work conditions, limited access to education and control over natural resources, including formal title to land, technology and credit, insecure employment and low income. This is due to a variety of factors, including the growing demand for flexible and cheap farm labor, the growing pressure on and conflicts over natural resources and the reallocation of economic resources in favor of large agroenterprises. Other factors include increasing exposure to risks related to natural disasters and environmental changes, worsening access to water, increasing occupational and health risks. Ongoing trends call for urgent actions in favor of gender and social equity in AKST policies and practices.

Women's Changing Forms of Involvement in Farm Activities and in the Management of Natural Resources

Women in agricultural production and postharvest activities range from 20 to 70%, and their involvement in farm activities, which is increasing in many developing countries, take on different and changing forms and statuses. Women's roles in agriculture varies in fact considerably according to farm system, legal systems, cultural norms and off-farm opportunities and are undergoing major transformations linked with local and global socioeconomic changes.

During a long period, women in industrialized countries either engaged in agricultural activities as farmers' spouses, or took off-farm employment. More recently the involvement of some women in farm activities has taken on a professional status as farm co-managers entitling them to pensions and other benefits of professional employment. Farm systems diversification and tertiarization have also favored the development of new economic activities taken up by women as autonomous entrepreneurs (direct sale, green tourism, etc.). In Central and Eastern European countries socialist policies historically aimed at suppressing gender differences in farm activities, a process that has been called into question by economic liberalization. Privatization of state and cooperatives farms resulted in fact in loss of employment for a large number of women. With EU integration however, countries (e.g., Poland) have benefited from EU support and training programs that also promoted new activities for rural women, such as on-farm processing, direct sale of farm products and agrotourism.

In certain industrialized countries (e.g., Spain, France) and in many developing regions, the consolidation of large export-oriented farm enterprises contributes to an increased number of female workers, including migrant workers in farm activities (e.g., horticulture, floriculture). This process of feminization of agricultural wage work is associated in some regions with the consolidation of large scale and export-oriented farm enterprises and the increasing demand of cheap labor. In developing countries it indicates the impoverishment of small farm households resulting in male out-migration to urban centers for work, and is also linked with rural women limited access to education and non-agricultural employment [CWANA Chapters 2; ESAP Chapter 1; Global Chapter 3].

In some countries (e.g., Tunisia, Morocco), progress in education has allowed more women to obtain university degrees or diplomas in agricultural sciences and to become farm entrepreneurs and managers. Still the proportion of female farm entrepreneurs remains very low in most developing countries (6% in Tunisia) and women's work is carried out on the basis of their status as family members, with little separation between domestic and productive activities.

Besides housekeeping and child rearing, women and girls are usually responsible for fetching water and fuel wood. Women and girls tend to perform tasks such as planting, transplanting, hand weeding, harvesting, picking fruit and vegetables, small livestock rearing, and postharvest operations such as threshing, seed selection, and storage, while mechanized work (preparing the land, irrigation, mechanical harvesting, and marketing) is generally a male task. This may increase women's and girls' manual and time burden, tends to keep girls out of school, and holds their productivity below their potential.

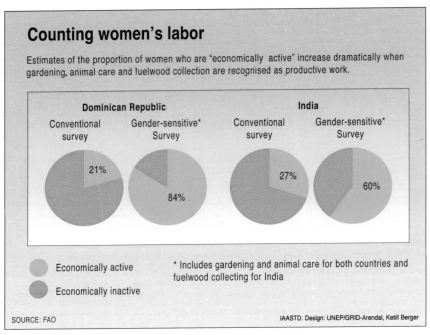

Counting women's labor

Estimates of the proportion of women who are "economically active" increase dramatically when gardening, animal care and fuelwood collection are recognised as productive work.

Dominican Republic — Conventional survey: 21% | Gender-sensitive* Survey: 84%

India — Conventional survey: 27% | Gender-sensitive* Survey: 60%

Economically active
Economically inactive

* Includes gardening and animal care for both countries and fuelwood collecting for India

SOURCE: FAO

IAASTD. Design: UNEP/GRID-Arendal, Ketill Berger

Figure SR-WA1. *Counting women's labor.*

As a result of male out-migration and the development of labor intensive farming systems, *the gender division in farm activities has undergone important transformation* and has tended to become more flexible. In some countries (e.g., in SSA) women are now in charge of tasks formerly performed only by men such as soil preparation, spraying and marketing. This requires women's access to additional skills and presents new risks (e.g., health risks related to the unregulated use of chemicals, especially pesticides) to girls and women.

Rural-to-urban *migration* and out-migration of men and young adults (including in some cases young women), especially in CWANA, ESAP, LAC and SSA regions, has increased the number of *female headed households* and has shifted the mean ages of rural populations upwards, resulting in considerable shrinkages in the rural labor force. In some cases, this has negatively affected agricultural production, food security, and service provision [Global Chapter 3]. As to decision-making, women in some cases have become empowered because of male out-migration: they manage budgets and their mobility is increased as they sometimes go to the market to sell their products, even if they still rely on male relatives for major decisions such as the sale of an animal (cow, veal, etc.) [CWANA Chapter 2; Global Chapter 6]. In Asia, SSA and LAC both internal and international migration by rural women seeking economic opportunities to escape poverty is on increase [ESAP Chapter 1].

Constraints, Challenges and Opportunities

The access of women to adequate land and land ownership continues to be limited due to legislation (e.g., Zimbabwe, Yemen) and sociocultural factors, e.g., Burundi where legislation has affirmed women's right to land but customary practices restrict women's ability to buy or inherit agricul-

tural land and resources [CWANA Chapter 1; SSA Chapter 2]. Agrarian reform programs tend to give title to men, especially in CWANA and LAC [CWANA Chapter 2; LAC Chapter 5]. In the majority of patrilineal societies, women's right to land expires automatically in the case of divorce or death of the husband [SSA Chapter 2]. In North Africa, inheritance law entitles women to half the amount endowed to men, and very often women forgo their right to land in favor of their brothers. Lack of control over and impaired entitlement to land often implies restricted access to loans and social security, limits autonomy and decision making power, and eventually curtails ability to achieve food security. A few countries have started recognizing the independent land rights of women (e.g., South Africa, Kenya) [Global Chapter 5; SSA Chapter 2]. The issue is the more urgent because market development rewards those who own the factors of production. Increased "opening toward the market" will not benefit men and women equally unless these institutional, legal and normative issues are appropriately and effectively addressed.

Poor rural infrastructure such as the lack of clean water supply, electricity or fuel increases women's work load and limits their availability for professional training, childcare and income generation. The lack of access to storage facilities and roads contributes to high food costs and low selling prices. The trends towards economic and *trade liberalization and privatization* have led to the dismantling of many marketing services that were previously available to farmers. Women farmers have been severely hit by this loss. The decline in investment in rural infrastructure, such as roads that link rural areas to markets and limited access to ICTs, affects women's access to markets. Lack of access to membership in marketing organizations limits women's ability to sell their produce.

Women and girls involved in farm activities mostly in

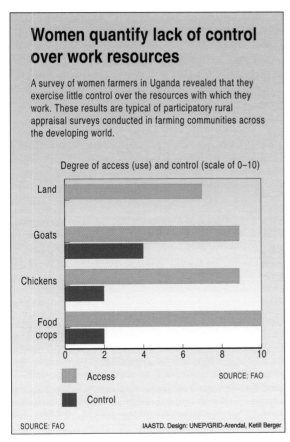

Figure SR-WA2. *Women quantify lack of control over work resources.*

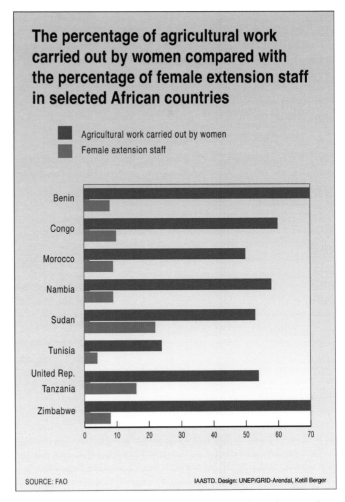

Figure SR-WA3. *The percentage of agricultural work carried out by women compared to the percentage of female extension staff in selected African countries.*

developing countries usually have less access than men *to education, information* and to learn how to use new technologies. Hence, this affects their ability to make informed choices around crop selection, food production and marketing. Notwithstanding a rise in the number of women pursuing careers in biosciences worldwide, female researchers still tend to be underrepresented in agricultural sciences and in senior scientific positions in general. Only 15% of the world's agricultural public sector *extension* agents are women [Global Chapter 3]. Women's access to extension is limited by lack of access to membership in rural organizations which often channel or provide training opportunities, and by gender blind agricultural policies that give inadequate attention to women farmer's needs in terms of crops and technology. Lack of opportunity in the curricula and training of extensionists to analyze gender roles and differential needs continues to exclude women from training and the benefits of extension services.

Although in most countries women have lower rates of *access to ICTs*, there are increasing examples of the use of ICTs by women to generate income (e.g., selling phone time in Bangladesh), obtain information, communicate with governments, and make their voices heard. In India, local women use video and radio equipment to record and produce the messages they want others in their community to hear (e.g., Deccan Development Society). The *Farmwise*

project in Malawi uses a computer database system with web interface and email to help women farmers determine what they can expect to harvest from their land, which crops they can grow given the soil type and fertility, and what inputs they should use [Global Chapter 6].

Access to information influences the ability of farmers to have influence in their communities and their ability to participate in *AKST decision-making*. Women's representation in AKST decision-making at all levels remains limited (e.g., women in Benin held 2.5% of high-level decision making positions in the government [Global Chapters 1, 2]. Women farmers' access to membership and leadership positions in rural organizations (e.g., cooperatives, agricultural producers' organizations, farmers' associations) is often restricted, by law or custom, which restricts their access to productive resources, credit, information and training and their ability to make their views known to policy makers and planners. The rise of women's Self-Help Groups (SHGs) or women's microfinance groups (e.g., in India) to some extent has made women's income a permanent component of household income, thus reducing women's dependency on the male provider [ESAP Chapter 5].

Although the supply of *gender disaggregated data and studies of women's roles in agricultural production* and food security is increasing, there is still a lack of sufficient data and in depth research on women's practices and specific needs. Indirect impacts of AKST in relation to ownership of assets, employment on and off farm, vulnerability, gender roles, labor requirements, food prices, nutrition and capacity for collective action have been less thoroughly researched than the financial and economic impacts, although, recent impact assessments of participatory methods have more comprehensively addressed these issues [Global Chapter 3].

Also agricultural research policies have tended to primarily focus on the intensive farming sector and export-oriented crops, and have given insufficient attention to food crops for domestic consumption, which are essential for household food security and environmental protection [Global Chapter 2]. Small-scale farmers, particularly women, play a key role in promoting sustainable methods of farming based on traditional knowledge and practices. Women often possess knowledge of the value and use of local plant and animal resources for nutrition, health and income in their roles as family caretakers, plant gatherers, home gardeners, herbalists, seed custodians and informal plant breeders [Global Chapter 2]. Moreover, women often experiment with and adapt indigenous species and thus become experts in plant genetic resources [SSA Chapter 2].

Climate change. Effects of flooding, drought, variations in crop seasons and temperature-related yield loss could mean extra hardship for the farming and food provisioning activities, which are often carried out by women. Their capacity to sustain their families' livelihoods is in fact often reduced as a result of the loss of seeds, livestock, tools and productive gardens [ESAP Chapter 4]. The increase of extreme weather conditions (e.g., floods and cyclones), notably in ESAP regions, will put an increasing expectation on women for coping with the effects of disaster and destruction.

Women are underrepresented in decision making about climate change, green house gas emissions and adaptation/mitigation in both the public and private sector. Lower levels of access to training, education and technologies will affect the ability of women to cope with climate change induced stresses.

Women of reproductive age as well as children are most affected by the increase of infectious *diseases* (e.g., malaria). The worsening health situation is exacerbated by a high rate of *malnutrition* in children especially in regions, like SSA, with repeated droughts, wars and conflicts. Intra-household food distribution often favors males, which can give rise to micronutrient deficiencies in women and children which impair cognitive development of young children, retard physical growth, increase child mortality and maternal death during childbirth [Global Chapter 3]. Nutritional deficiency among women and children in South Asia also has reached crisis proportions [ESAP Chapter 1]. The impact of *HIV/AIDS* in an increasing number of countries has given rise to rapidly increasing numbers of female-headed households, child-headed households, and dependence on the elderly who face increasing workloads as they assume responsibil-

ity for growing numbers of AIDS orphans [SSA Chapter 3]. In SSA women make up two-thirds of those infected with HIV/AIDS. This adds additional burdens for women as producers of food and as family caretakers. Labor loss due to illness, need to care for family members and paid employment required to cover medical costs may cause families to decrease their farming activities The stress of HIV/AIDS on the social capital within communities also erodes the transmission of knowledge between households and communities, thereby reducing the range of livelihood options for the next generation [Global Chapters 6, 7].

Options for Action to Enhance Women's Involvement in AKST
In view of the continuing constraints faced by rural women and the current forms of agricultural restructuring likely to worsen farm women's work and health conditions, urgent action is needed to implement gender and social equity in AKST policies and practices.

Options for action include:
- Strengthening the capacity of public institutions and NGOs to improve the knowledge of women's involvement in farm activities and their relationship to AKST;
- Giving priority to women's access to education, information, science and technology and extension services;
- Improving women's access, ownership and control of economic and natural resources through legal measures, appropriate credit schemes, support to the development of women's income generating activities and the reinforcement of women's organizations and networks;

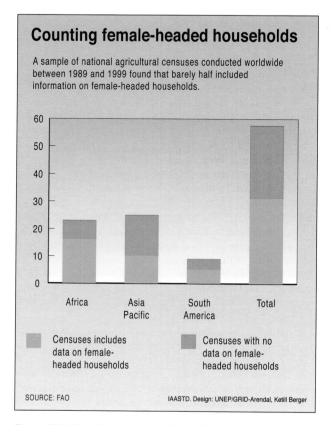

Counting female-headed households

A sample of national agricultural censuses conducted worldwide between 1989 and 1999 found that barely half included information on female-headed households.

Legend:
- Censuses includes data on female-headed households
- Censuses with no data on female-headed households

SOURCE: FAO

IAASTD. Design: UNEP/GRID-Arendal, Ketill Berger

Figure SR-WA4. *Counting female-headed households.*

- Strengthening women's ability to benefit from market-based opportunities by market institutions and policies giving explicit priority to women farmers groups in value chains;
- Supporting public services and investment in rural areas in order to improve women's living and working conditions;
- Prioritizing technological development policies targeting rural and farm women's needs and recognizing women's specific knowledge, skills and experience in the production of food and the conservation of biodiversity;
- Assessing the effects of farming practices and technology, including pesticides on women's health, and measures to reduce use and exposure;
- Ensuring gender balance in AKST decision-making at all levels; and
- Providing mechanisms to hold AKST organizations accountable for progress in the above areas.

Policies can reinforce the achievement of development and sustainability goals by *recognizing and taking into account the role played by family farming and rural women* in terms of production, employment and household food sufficiency. Consolidation of the small-scale farming sector, where women are particularly active, requires AKST oriented towards the improvement of local food crops to better satisfy domestic markets, the development of drought-resistant breeds to provide a more reliable harvest to those living on marginal lands, and greater focus on on-farm enterprises such as seasonal fish ponds that increase women's economic contribution to household survival.

Strengthening women's control over resources is central to achievement of development and sustainability goals as well as *changes in discriminatory laws* that exclude women from land ownership, from access to clean water, getting loans or opening bank accounts. The principle of equal pay for women working in agriculture, innovative low-cost and sustainable technological options and services in water supply are among the measures that can enable more equitable benefit-sharing from AKST investments and wider access to services that benefit both women and men. Governments can facilitate access to grants or credit on concessionary terms to women and women's groups.

There is an urgent need for *priority setting in research* to ensure that women benefit from modern agricultural technologies (e.g., labor-saving technologies and reduced health risk techniques) rather than being overlooked in the implementation of technologies as has often occurred in the past [Global Chapter 3]. For social and economic sustainability, it is important that technologies are appropriate to different resource levels, including those of women and do not encourage others to dispossess women of land or control their labor and income. Development of techniques that reduce work load and health risks, and meet the social and physical requirements of women can contribute to limiting the negative effects of the gender division of labor in many regions.

Modern agricultural technology should not undermine women's autonomy and economic position. Targeted measures will be needed to ensure this does not happen. AKST systems that are gender sensitive would expand the range of crop, horticultural, medicinal and animal species and varieties available for food provisioning and market sale. They would take into account all phases of agronomic management and post-harvest activities. Policy makers and researchers would need to consider the complex social, health and environmental implications of adopting engineered crops and weigh these against lost opportunities to direct institutional attention towards proven low external input agroecological approaches and strengthening farmer-centered seed-saving networks. By integrating local and gender-differentiated understanding of seeds and the cultural values connected to food preservation, preparation and storage, AKST can enhance the success of technological adoption and eventually be more effective in enhancing rural livelihoods.

Intellectual Property Rights that recognize women's technological knowledge and biological materials are needed if development and sustainability goals are to be met. Women's intellectual property rights relating to the knowledge of indigenous plant varieties and cultivation are in need of protection. Support of the documentation and dissemination of women's knowledge is an important aspect of a gender-sensitive approach to IPR [Global Chapter 2] and is required to retain the knowledge of both women and men.

As disaster-related and complex emergencies will become more frequent and larger in scale, preferential *research aiming at a better understanding of how gender issues affect communities' vulnerability* and their ability to respond is indispensable. Gender differences in vulnerability and in adaptive opportunities should be better researched and acknowledged in the technology development to mitigate carbon emissions ensuring success of adaptation policies.

Communities and civil society could be further supported to voice their concern for *gender-sensitive agricultural services*. They could assist in *collecting information* on men and women's roles, access, needs of AKST in different societies (including nomadic communities) and in sharing this on broader platforms, in order to have gender issues taken seriously in the design of development plans and agricultural services. Agricultural programs designed to increase women's income and household nutrition would need to take much greater account of the cultural context of women's work as well as patterns of intra-household food distribution and natural resource access if development and sustainability goals are to be met [Global Chapter 3].

Giving preference and support *women's access to education and information* is critical to meeting development and sustainability goals. Targeting female students for advanced education in agriculture and other sciences is a vital part of this preference as well as curriculum reform that expands the scope of knowledge relevant to meeting development and sustainability goals. This priority should be placed in the larger social, environmental or "life" context: the Earth University in Costa Rica combines hands-on fieldwork experience with theoretical work on not only the agricultural sciences, but also business administration, entrepreneurship, ecology, resource management, forestry, anthropology and sociology.

Training women farmers as trainers for other women provides an opportunity to share their experience and knowledge. Training and micro-credit programs should be interlinked to effectively transfer agricultural technology to women farmers. Marketing, food processing and post-harvest sciences are well suited as areas of specialization for women who desire a career in extension work. Strategies can include *making extension work attractive to women* and promoting the education and hiring of women as extension agents. Relevant expertise includes improved postharvest handling practices in the local marketplaces where women gather to sell their goods or to shop for food [Global Chapter 6].

Gender-sensitive communication strategies for natural resource management (e.g., mountain landscapes, trees-outside-forest, forest management) can ensure that women and girls can participate effectively and equitably in emerging knowledge networks. The availability of women-oriented content and selection of appropriate intermediaries and partnerships can enhance womens' and girls' access to and benefits from modern ICTs [Global Chapter 5]. Other benefits of ICT include linking up training and micro-credit programs to transfer agricultural technology between women farmers. Linking women farmers with markets and using effective, appropriate and cost-efficient ICTs can promote skills development among women. The use of the mobile phone is an example of an information technology that is increasing exponentially among women in many developing regions. Mobile phones are also a portable market research tool, allowing producers to find and compare current market prices for their products and ensuring greater profits for their products [Global Chapters 2, 6].

Furthering gender analysis in the alternative trade sector, particularly by Fair Trade organizations and NGOs, would generate a richer understanding of the costs and benefits in participating in alternative trade systems for both women and men. Gender impact analyses in turn can inform producer organizations and alternative trade organizations on how to improve their impact and on whom to focus further capacity development efforts. Such findings might point for instance to the need for female extension agents, or gender specific technology, marketing strategies or knowledge for male or female farmers.

Strengthening women's ability to benefit from market-based opportunities by market institutions and policies giving explicit *priority to women farmers groups in value chains* is essential and would allow women to benefit more from the added value of agricultural production. The development of agricultural enterprises owned and controlled by women, promoting women's organizations and cooperatives, community-supported agriculture and farmers markets have proven potential to enhance women's income opportunities and business capacities.

Strengthening *women's participation in formal AKST decision-making at all levels*, including international agricultural research centers and national agricultural research systems, is of crucial importance. Specific mechanisms should also be developed to hold AKST organizations accountable for progress in the above areas. Adoption of techniques such as gender budgeting by departments/programs of agriculture would assist in the allocation of public and private investments needed to implement (and assess) gender and social equity in AKST policies.

Annex A
Reservations on Synthesis Report

Australia: Australia recognizes the IAASTD initiative and reports as a timely and important multistakeholder and multidisciplinary exercise designed to assess and enhance the role of AKST in meeting the global development challenges. The wide range of observations and views presented however, are such that Australia cannot agree with all assertions and options in the report. The report is therefore noted as a useful contribution, which will be used for considering the future priorities and scope of AKST in securing economic growth and the alleviation of hunger and poverty.

Canada: In recognizing the important and significant work undertaken by IAASTD authors, Secretariat and stakeholders on the background Reports, the Canadian Government notes these documents as a valuable and important contribution to policy debate which needs to continue in national and international processes. While acknowledging the valuable contribution these Reports provide to our understanding on agricultural knowledge, science and technology for development, there remain numerous areas of concern in terms of balanced presentation, policy suggestions and other assertions and ambiguities. Nonetheless, the Canadian Government advocates these reports be drawn to the attention of governments for consideration in addressing the importance of AKST and its large potential to contribute to economic growth and the reduction of hunger and poverty.

United States of America: The United States joins consensus with other governments in the critical importance of AKST to meet the goals of the IAASTD. We commend the tireless efforts of the authors, editors, Co-Chairs and the Secretariat. We welcome the IAASTD for bringing together the widest array of stakeholders for the first time in an initiative of this magnitude. We respect the wide diversity of views and healthy debate that took place.

As we have specific and substantive concerns in each of the reports, the United States is unable to provide unqualified endorsement of the reports, and we have noted them.

The United States believes the Assessment has potential for stimulating further deliberation and research. Further, we acknowledge the reports are a useful contribution for consideration by governments of the role of AKST in raising sustainable economic growth and alleviating hunger and poverty.

Authors and Review Editors of Global and Sub-Global Reports

Argentina
Walter Ismael Abedini • Universidad Nacional de La Plata
Hugo Cetrángolo • Universidad de Buenos Aires
Cecilia Gelabert • Universidad de Buenos Aires
Héctor D. Ginzo • Ministerio de Relaciones Exteriores, Comercio Internacional y Culto
Maria Cristina Plencovich • Universidad de Buenos Aires
Marcelo Regunaga • Universidad de Buenos Aires
Sandra Elizabeth Sharry • Universidad Nacional de La Plata
Javier Souza Casadinho • CETAAR-RAPAL
Miguel Taboada • Universidad de Buenos Aires
Ernesto Viglizzo • INTA Centro Regional La Pampa

Armenia
Ashot Hovhannisian • Ministry of Agriculture

Australia
Helal Ahammad • Department of Agriculture, Fisheries and Forestry
David J. Connor • University of Melbourne
Tony Jansen • TerraCircle Inc.
Roger R.B. Leakey • James Cook University
Andrew Lowe • Adelaide State Herbarium and Biosurvey
Anna Matysek • Concept Economics
Andrew Mears • Majority World Technology
Girija Shrestha • Monash Asia Institute, Monash University

Austria
Maria Wurzinger • University of Natural Resources & Applied Life Sciences

Bangladesh
Wais Kabir • Bangladesh Agricultural Research Council (BARC)
Karim Mahmudul • Bangladesh Shrimp and Fish Foundation

Barbados
Carl B. Greenidge • CFTC and Caribbean Regional Negotiating Machinery

Benin
Peter Neuenschwander • International Institute of Tropical Agriculture
Simplice Davo Vodouhe • Pesticide Action Network

Bolivia
Jorge Blajos • PROINPA Foundation
Ruth Pamela Cartagena • CIPCA Pando
Manuel de la Fuente • National Centre of Competence in Research North-South
Edson Gandarillas • PROINPA Foundation

Botswana
Baone Cynthia Kwerepe • Botswana College of Agriculture

Brazil
Flavio Dias Ávila • Embrapa
Antônio Gomes de Castro • Embrapa
André Gonçalves • Centro Ecológico
Dalva María Da Mota • Embrapa
Odo Primavesi • Embrapa Pecuaria Sudeste (Southeast Embrapa Cattle)
Sergio Salles Filho • State University of Campinas (Unicamp)
Susana Valle Lima • Embrapa

Canada
Jacqueline Alder • University of British Columbia
Guy Debailleul • Laval University
Harriet Friedman • University of Toronto
Tirso Gonzales • University of British Columbia, Okanagan
Thora Martina Herrmann • Université de Montréal
Sophia Huyer • UN Commission on Science and Technology for Development.
JoAnn Jaffe • University of Regina
Shawn McGuire
Morven A. McLean • Agriculture and Biotechnology Strategies Inc. (AGBIOS)
M. Monirul Qader Mirza • Environment Canada and University of Toronto, Scarborough
Ricardo Ramirez • University of Guelph
John M.R. Stone • Carleton University

Chile
Mario Ahumada • International Committee for Regional Planning for Food Security

China
Jikun Huang • Chinese Academy of Sciences
Fu Quin • Chinese Academy of Agricultural Sciences (CAAS)
Ma Shiming • Chinese Academy of Agricultural Sciences (CAAS)
Li Xiande • Chinese Academy of Agricultural Sciences (CAAS)
Zhu Xiaoman • China National Institute for Educational Research

Colombia

Inge Armbrecht • University del Valle
Hernando Bernal • University of the Columbian Amazon
Juan Cárdenas • University of the Andes
Maria Veronica Gottret • CIAT
Elsa Nivia • RAPALMIRA
Edelmira Pérez • Pontificia University Javeriana of Bogotá

Costa Rica

Marian Perez Gutierrez • National Centre of Competence in
    Research North-South
Mario Samper • Inter-American Institute for Cooperation on
    Agriculture (IICA)

Côte d'Ivoire

Guéladio Cissé • National Centre of Competence in Research
    North-South Centre Suisse de Recherche Scientifique

Cyprus

Georges Eliades • Agricultural Research Institute (ARI)
Costas Gregoriou • Agricultural Research Institute (ARI)
Christoph Metochis • Agricultural Research Institute (ARI)

Czech Republic

Miloslava Navrátilová • State Phytosanitary Administration

Democratic Republic of Congo

Dieudonne Athanase Musibono • University of Kinshasa

Denmark

Henrik Egelyng • Danish Institute for International Studies (DIIS)
Thomas Henrichs • University of Aarhus

Dominican Republic

Rufino Pérez-Brennan • ALIMENTEC S.A.

Egypt

Sonia Ali • Zagarid University
Mostafa A. Bedier • Agricultural Economic Research Institute
Salwa Mohamed Ali Dogheim • Agriculture Research Center
Azza Emara • Agricultural Research Institute, Agricultural
    Research Center
Ahmed Abd Alwahed Rafea • American University of Cairo
Mohamed Abo El Wafa Gad • GTZ

Ethiopia

Assefa Admassie • Ethiopian Economic Policy Research Institute
P. Anandajayasekeram • International Livestock Research
    Institute
Gezahegn Ayele • EDRI-IFPRI
Berhanu Debele • National Centre of Competence in Research
    North-South
Joan Kagwanja • Economic Commission for Africa
Yalemtsehay Mekonnen • Addis Ababa University
Workneh Negatu Sentayehu • Addis Ababa University
Gete Zeleke • Global Mountain Program

Finland

Riina Antikainen • Finnish Environment Institute
Henrik Bruun • Helsinki University of Technology
Helena Kahiluoto • MTT Agrifood Research

Jyrki Niemi • MTT Agrifood Research
Riikka Rajalahti • Ministry of Foreign Affairs
Reimund Roetter • MTT Agrifood Research
Timo Sipiläinen • MTT Agrifood Research
Markku Yli-Halla • University of Helsinki

France

Jean Albergel • Institut National de la Recherche Agronomique
    (INRA)
Loïc Antoine • IFREMER
Martine Antona • CIRAD
Gilles Aumont • Institut National de la Recherche Agronomique
    (INRA)
Didier Bazile • CIRAD
Pascal Bergeret • Ministry of Agriculture
Yves Birot • Institut National de la Recherche Agronomique
    (INRA)
Pierre-Marie Bosc • CIRAD
Nicolas Bricas • CIRAD
Jacques Brossier • Institut National de la Recherche.
    Agronomique (INRA)
Perrine Burnod • CIRAD
Gérard Buttoud • Institut National de la Recherche Agronomique
    (INRA)
Patrick Caron • CIRAD
Bernard Chevassus • French Ministry of Agriculture and Fisheries
Emilie Coudel • CIRAD
Béatrice Darcy-Vrillon • Institut National de la Recherche
    Agronomique (INRA)
Jean-François Dhôte • Institut National de la Recherche
    Agronomique (INRA)
Celine Dutilly-Diane • CIRAD
Fabrice Dreyfus • University Institute for Tropical Agrofood
    Industries and Rural Development
Michel Dulcire • CIRAD
Patrick Dugué • CIRAD
Nicolas Faysse • CIRAD
Stefano Farolfi • CIRAD
Guy Faure • CIRAD
Alia Gana • National Center for Scientific Research CNRS/
    LADYSS
Thierry Goli • CIRAD
Ghislain Gosse • Institut National de la Recherche Agronomique
    (INRA)
Jean-Marc Guehl • Institut National de la Recherche
    Agronomique (INRA)
Dominique Hervé • Institute for Development Research (IRD)
Henri Hocdé • CIRAD
Bernard Hubert • Institut National de la Recherche Agronomique
    (INRA)
Jacques Imbernon • CIRAD
Hugues de Jouvenel • Futuribles
Trish Kammili • Institut National de la Recherche Agronomique
Véronique Lamblin • Futuribles
Marie de Lattre-Gasquet • CIRAD
Patrick Lavelle • Institute for Development Research (IRD)
Marianne Lefort • Institut National de la Recherche Agronomique
    and AgroParisTech
Jacques Loyat • French Ministry of Agriculture and Fisheries
Jean-Pierre Müller • CIRAD
Sylvain Perret • CIRAD

Michel Petit • Institut Agronomique Mediterraneen Montpellier
Jean-Luc Peyron • GIP ECOFOR
Anne-Lucie Raoult-Wack • Agropolis Fondation
Pierre Ricci • Institut National de la Recherche Agronomique
    (INRA)
Alain Ruellan • Agrocampus Rennes
Yves Savidan • AGROPOLIS
Bernard Seguin • Institut National de la Recherche Agronomique
    (INRA)
Nicole Sibelet • CIRAD
Andrée Sontot • Bureau de Ressources Genetiques
Ludovic Temple • CIRAD
Jean-Philippe Tonneau • CIRAD
Selma Tozanli • Mediterranean Agronomic Institute of Montpellier
Guy Trebuil • CIRAD
Tancrede Voituriez • CIRAD

The Gambia
Ndey Sireng Bakurin • National Environment Agency

Germany
Anita Idel • Independent
Dale Wen Jiajun • International Forum on Globalization
Tanja H. Schuler • Independent
Hermann Waibel • Leibniz University of Hannover

Ghana
Elizabeth Acheampong • University of Ghana
John-Eudes Andivi Bakang • Kwame Nkrumah University of
    Science and Technology (KNUST)
Claudio Bragantini • Embrapa
Daniel N. Dalohoun • United Nations University MERIT/INRA
Felix Yao Mensa Fiadjoe • University of Ghana
Edwin A. Gyasi • University of Ghana
Gordana Kranjac-Berisavljevic • University for Development
    Studies
Carol Mercey Markwei • University of Ghana Legon
Joseph (Joe) Taabazuing • Ghana Institute of Management and
    Public Administration (GIMPA)

India
Satinder Bajaj • Eastern Institute for Integrated Learning in
    Management University
Sachin Chaturvedi • Research and Information System for
    Developing Countries (RIS)
Indu Grover • CCS Haryana Agricultural University
Govind Kelkar • UNIFEM
Purvi Mehta-Bhatt • Science Ashram
Poonam Munjal • CRISIL Ltd
Dev Nathan • Institute for Human Development
K.P. Palanisami • Tamil Nadu Agricultural University
Rajeswari Sarala Raina • Centre for Policy Research
Vanaja Ramprasad • Green Foundation
C.R. Ranganathan • Tamil Nadu Agricultural University
Sunil Ray • Institute of Development Studies
Sukhpal Singh • Indian Institute of Management (IIM)
Anushree Sinha • National Council for Applied Economic
    Research (NCAER)
V. Santhakumar • Centre for Development Studies
Rasheed Sulaiman V. • Centre for Research on Innovation and
    Science Policy (CRISP)

Indonesia
Suraya Afiff • KARSA (Circle for Agrarian and Village Reform)
Hira Jhamtani • Third World Network

Iran
Hamid Siadat • Independent

Ireland
Denis Lucey • University College Cork – National University of
    Ireland

Italy
Gustavo Best • Independent
Maria Fonte • University of Naples
Michael Halewood • Bioversity International
Anne-Marie Izac • Alliance of the CGIAR Centres
Prabhu Pingali • FAO
Sergio Ulgiati • Parthenope University of Naples
Francesco Vanni • Pisa University
Keith Wiebe • FAO
Monika Zurek • FAO

Jamaica
Audia Barnett • Scientific Research Council

Japan
Osamu Ito • Japan International Research Center for Agricultural
    Sciences (JIRCAS)
Osamu Koyama • Japan International Research Center for
    Agricultural Sciences (JIRCAS)

Jordan
Saad M. Alayyash • Jordan University of Science and Technology
Ruba Al-Zubi • Ministry of Environment
Mahmud Duwayri • University of Jordan
Muna Yacoub Hindiyeh • Jordan University of Science and
    Technology
Lubna Qaryouti • Ministry of Agriculture/Rangelands Directorate
Rania Suleiman Shatnawi • Ministry of Environment

Kenya
Tsedeke Abate • International Crops Research Institute for the
    Semi-Arid Tropics
Susan Kaaria • Ford Foundation
Boniface Kiteme • Centre for Training and Integrated Research in
    Arid and Semi-arid Lands Development
Washington O. Ochola • Egerton University
Wellington Otieno • Maseno University
Frank M. Place • World Agroforestry Centre
Wahida Patwa Shah • ICRAF – World Agroforestry Centre

Kyrgyz Republic
Ulan Kasymov • Central Asian Mountain Partnership Programme
Rafael Litvak • Research Institute of Irrigation

Latvia
Rashal Isaak • University of Latvia

Lebanon
Roy Antoine Abijaoude • Holy Spirit University

Madascagar
R. Xavier Rakotonjanahary • FOFIFA (National Center for Applied Research for Rural Development)

Malaysia
Lim Li Ching • Third World Network
Khoo Gaik Hong • International Tropical Fruits Network

Mauritius
Ameenah Gurib-Fakim • University of Mauritius

Mexico
Rosa Luz González Aguirre • Autonomous Metropolitan University, Azcapotzalco
Michelle Chauvet • Autonomous National University of México (UNAM)
Amanda Gálvez • Autonomous National University of México (UNAM)
Jesús Moncada • Independent
Celso Garrido Noguera • Autonomous National University of México (UNAM)
Scott S. Robinson • Universidad Metropolitana - Iztapalapa
Roberto Saldaña • SAGARPA

Morocco
Saadia Lhaloui • Institut National de la Recherche Agronomique
Mohamed Moussaoui • Independent

Mozambique
Manuel Amane • Instituto de Investigação Agrícola de Moçambique (IIAM)
Patrick Matakala • World Agroforestry Centre

Nepal
Rajendra Shrestha • AFORDA

Netherlands
Nienke Beintema • International Food Policy Research Institute
Bas Eickhout • Netherlands Environmental Assessment Agency (MNP)
Judith Francis • Technical Centre for Agricultural and Rural Cooperation (CTA)
Janice Jiggins • Wageningen University
Toby Kiers • Vrije Universiteit
Kaspar Kok • Wageningen University
Niek Koning • Wageningen University
Niels Louwaars • Wageningen University
Willem A. Rienks • Wageningen University
Niels Röling • Wageningen University
Mark van Oorschot • Netherlands Environmental Assessment Agency (MNP)
Detlef P. van Vuuren • Netherlands Environmental Assessment Agency (MNP)
Henk Westhoek • Netherlands Environmental Assessment Agency (MNP)

New Zealand
Jack A. Heinemann • University of Canterbury
Meriel Watts • Pesticide Action Network Aotearoa

Nicaragua
Falguni Guharay • Information Service of Mesoamerica on Sustainable Agriculture
Carlos J. Pérez • Earth Institute
Ana Cristina Rostrán • UNAN-León
Jorge Irán Vásquez • National Union of Farmers and Ranchers

Nigeria
Sanni Adunni • Ahmadu Bello University
Michael Chidozie Dike • Ahmadu Bello University
V.I.O. Ndirika • Ahmadu Bello University
Stella Williams • Obafemi Awolowo University

Oman
Younis Al Akhzami • Ministry of Agriculture and Fisheries
Abdallah Mohamed Omezzine • University of Nizwa, Oman

Pakistan
Iftikhar Ahmad • National Agricultural Research Centre
Mukhtar Ahmad Ali • Centre for Peace & Development Initiatives
Syed Sajidin Hussain • Ministry of Environment
Yameen Memon • Government Employees Cooperative Housing Society
Farzana Panhwar • SINDTH Rural Women's Uplift Group
Syed Wajid Pirzada • Pakistan Agricultural Research Center
Abid Suleri • Sustainable Development Policy Institute (SDPI)
Ahsan Wagha • Damaan Development Organization/GEF/SGP

Palestine
Jamal Abo Omar • An-Najah National University
Jad E Isaac • Applied Research Institute – Jerusalem
Thameen Hijawi • Palestinian Agricultural Relief Committees (PARC)
Numan Mizyed • An-Najah National University
Azzam Saleh • Al-Quds University

Panama
Julio Santamaría • INIAP

Peru
Clara G. Cruzalegui • Ministry of Agriculture and Livestock
Maria E. Fernandez • National Agrarian University
Luis A. Gomero • Action Network for Alternatives to Agrochemicals
Carla Tamagno • Universidad San Martin de Porres

Philippines
Mahfuz Ahmed • Asian Development Bank
Arturo S. Arganosa • Philippine Council for Agriculture, Forestry and Natural Resources Research and Development
Danilo C. Cardenas • Philippine Council for Agriculture, Forestry and Natural Resources Research and Development
Richard B. Daite • Philippine Council for Agriculture, Forestry and Natural Resources Research and Development
Elenita C. Dano • Participatory Enhancement and Development of Genetic Resources in Asia (PEDIGREA)
Fezoil Luz C. Decena • Philippine Council for Agriculture, Forestry and Natural Resources Research and Development
Dely Pascual Gapasin • Institute for International Development Partnership Foundation

Digna Manzanilla • Philippine Council for Agriculture, Forestry and Natural Resources Research and Development
Charito P. Medina • MASIPAG (Farmer-Scientist Partnership for Development, Inc.)
Thelma Paris • International Rice Research Institute
Agnes Rola • University of the Philippines Los Baños
Leo Sebastian • Philippine Rice Research Institute

Poland
Dariusz Jacek Szwed • Independent
Dorota Metera • IUCN – Poland

Russia
Sergey Alexanian • N.I. Vavilov Research Institute of Plant Industry

Rwanda
Agnes Abera Kalibata • Ministry of Agriculture

Senegal
Julienne Kuiseu • CORAF/WECARD
Moctar Toure • Independent

Slovakia
Pavol Bielek • Soil Science and Conservation Research Institute

South Africa
Urmilla Bob • University of KwaZulu-Natal
Marnus Gouse • University of Pretoria
Moraka Makhura • Development Bank of Southern Africa

Spain
Maria del Mar Delgado • University of Córdoba
Mario Giampietro • Universitat Autònoma de Barcelona
Luciano Mateos • Instituto de Agricultura Sostenible, CSIC
Marta Rivera-Ferre • Autonomous University of Barcelona

Sri Lanka
Deborah Bossio • International Water Management Institute
Charlotte de Fraiture • International Water Management Institute
Francis Ndegwa Gichuki • International Water Management Institute
David Molden • International Water Management Institute

Sudan
Ali Taha Ayoub • Ahfal University for Women
Asha El Karib • ACORD
Aggrey Majok • Independent
Ahmed S.M. El Wakeel • NBSAP
Balgis M.E. Osman-Elasha • Higher Council for Environment & Natural Resources (HCENR)

Sweden
Susanne Johansson • Swedish University of Agricultural Sciences
Richard Langlais • Nordregio, Nordic Center for Spatial Devleopment
Veli-Matti Loiske • Södertörns University College
Fred Saunders • Södertörns University College
Martin Wierup • Swedish University of Agricultural Sciences

Switzerland
Felix Bachmann • Swiss College of Agriculture

David Duthie • United Nations Environment Programme
Markus Giger • University of Bern
Ann D. Herbert • International Labour Organization
Angelika Hilbeck • Swiss Federal Institute of Technology
Udo Hoeggel • University of Bern
Hans Hurni • University of Bern
Andreas Klaey • University of Bern
Cordula Ott • University of Bern
Brigitte Portner • University of Bern
Stephan Rist • University of Bern
Urs Scheidegger • Swiss College of Agriculture
Juerg Schneider • State Secretariat for Economic Affairs
Christoph Studer • Swiss College of Agriculture
Hong Yang • Swiss Federal Institute for Aquatic Science and Technology.
Yuan Zhou • Swiss Federal Institute for Aquatic Science and Technology
Christine Zundel • Research Institute of Organic Agriculture (FiBL)

Syria
Nour Chachaty • Independent
Alessandra Galie • ICARDA
Stefania Grando • ICARDA
Theib Yousef Oweis • ICARDA
Manzoor Qadir • ICARDA
Kamil H. Shideed • ICARDA

Taiwan
Mubarik Ali • World Vegetable Center

Tajikistan
Sanginov S. Rajabovich • Soil Science Research Institute of Agrarian Academy of Sciences

Tanzania
Roshan Abdallah • Tropical Pesticides Research Institute (TPRI)
Stella N. Bitende • Ministry of Livestock and Fisheries Development
Sachin Das • Animal Diseases Research Institute
Aida Cuthbert Isinika • Sokoine University of Agriculture
Rose Rita Kingamkono • Tanzania Commission for Science & Technology
Evelyne Lazaro • Sokoine University of Agriculture
Razack Lokina • University of Dar es Salaam
Lutgard Kokulinda Kagaruki • Animal Diseases Research Institute
Elizabeth J.Z. Robinson • University of Dar es Salaam

Thailand
Thammarat Koottatep • Asian Institute of Technology
Anna Stabrawa • United Nations Environment Programme

Trinidad and Tobago
Salisha Bellamy • Ministry of Agriculture, Land & Marine Resources
Ericka Prentice-Pierre • Agriculture Sector Reform Program (ASRP), IBD

Tunisia
Mohamed Annabi • Institut National de la Recherche Agronomique de Tunisie

Rym Ben Zid • Independent
Mustapha Guellouz • IAASTD CWANA, DSIPS - Diversification Program, ICARDA
Kawther Latiri • Institut National de la Recherche Agronomique de Tunisie
Lokman Zaibet • Ecole Supérieure d'Agriculture de Mograne, Zaghouan

Turkey
Nazimi Acikgoz • Ege University
Hasan Akca • Gaziosmanpasa University
Ahmet Ali Koc • Akdeniz University
Gulcan Eraktan • University of Ankara
Yalcin Kaya • Trakya Agricultural Research Institute
Suat Oksuz • Ege University
Ayfer Tan • Aegean Agricultural Research Institute
Ahu UncuogluTubitak • Research Institute for Genetic Engineering and Biotechnology (RIGEB)
Fahri Yavuz • Ataturk University

Uganda
Apili E.C. Ejupu • Ministry of Agriculture, Animal Industries and Fisheries
Apophia Atukunda • Environment Consultancy League
Dan Nkoowa Kisauzi • Nkoola Institutional Development Associates (NIDA)
Imelda Kashaija • National Agriculture Resource Organization (NARO)
Theresa Sengooba • International Food Policy Research Institute

Ukraine
Yuriy Nesterov • Heifer International

United Arab Emirates
Abdin Zein El-Abdin • Lootah Educational Foundation

United Kingdom
Michael Appleby • World Society for the Protection of Animals, London
Steve Bass • International Institute for Environment and Development
Stephen Biggs • University of East Anglia
Norman Clark • The Open University
Joanna Chataway • Open University
Janet Cotter • University of Exeter
Peter Craufurd • University of Reading
Barbara Dinham • Pesticide Action Network
Cathy Rozel Farnworth • Independent
Les Firbank • North Wyke Research
Chris Garforth • University of Reading
Anil Graves • Cranfield University
Andrea Grundy • National Farmers' Union
David Grzywacz • University of Greenwich
Andy Hall • United Nations University – Maastricht
Brian Johnson • Independent
Sajid Kazmi • Middlesex University Business School
Frances Kimmins • NR International Ltd
Chris D.B. Leakey • University of Plymouth
Karen Lock • London School of Hygiene and Tropical Medicine
Peter Lutman • Rothamsted Research
Ana Marr • University of Greenwich

John Marsh • Independent
Adrienne Martin • University of Greenwich
Ian Maudlin • Centre for Tropical Veterinary Medicine
Nigel Maxted • University of Birmingham
Mara Miele • Cardiff University
Selyf Morgan • Cardiff University
Joe Morris • Cranfield University
Johanna Pennarz • ITAD
Gerard Porter • University of Edinburgh
Charlie Riches • University of Greenwich
Peter Robbins • Independent
Paresh Shah • London Higher
Geoff Simm • Scottish Agricultural College
Linda Smith • Department for Environment, Food and Rural Affairs (end Mar 2006)
Nicola Spence • Central Science Laboratory
Joyce Tait • University of Edinburgh
K.J. Thomson • University of Aberdeen
Philip Thornton • International Livestock Research Institute
Bill Vorley • International Institute for Environment and Development
Jeff Waage • London International Development Centre

United States
Emily Adams • Independent
Elizabeth A. Ainsworth • U.S. Department of Agriculture
Wisdom Akpalu • Environmental Economics Research & Consultancy (EERAC)
Molly D. Anderson • Food Systems Integrity
David Andow • University of Minnesota
Patrick Avato • The World Bank
Mohamed Bakarr • Center for Applied Biodiversity Science, Conservation International
Revathi Balakrishnan • Independent
Debbie Barker • International Forum on Globalization
Barbara Best • U.S. Agency for International Development
Regina Birner • International Food Policy Research Policy Institute
Dave Bjorneberg • U.S. Department of Agriculture
David Bouldin • Cornell University
Rodney Brown • Brigham Young University
Sandra Brown • Winrock International
Rebecca Burt • U.S. Department of Agriculture
Lorna M. Butler • Iowa State University
Kenneth Cassman • University of Nebraska, Lincoln
Gina Castillo • Oxfam America
Medha Chandra • Pesticide Action Network, North America
Jahi Michael Chappell • University of Michigan
Luis Fernando Chávez • Emory University
Joel I. Cohen • Independent
Randy L. Davis • U.S. Department of Agriculture
Daniel de la Torre Ugarte • University of Tennessee
Steven Dehmer • University of Minnesota
Medha Devare • Cornell University
Amadou Makhtar Diop • Rodale Institute
William E. Easterling • Pennsylvania State University
Kristie L. Ebi • ESS, LLC
Denis Ebodaghe • U.S. Department of Agriculture
Shelley Feldman • Cornell University
Shaun Ferris • Catholic Relief Services
Jorge M. Fonseca • University of Arizona

J.B. Friday • University of Hawaii
Tilly Gaillard • Independent
Constance Gewa • George Mason University
Paul Guillebeau • University of Georgia
James C. Hanson • University of Maryland
Celia Harvey • Conservation International
Mary Hendrickson • University of Missouri
William Heffernan • University of Missouri
Paul Heisey • U.S. Department of Agriculture
Kenneth Hinga • U.S. Department of Agriculture
Omololu John Idowu • Cornell University
Marcia Ishii-Eiteman • Pesticide Action Network, North America
R. Cesar Izaurralde • Joint Global Change Research Institute
Eric Holt Jiménez • Food First/Institute for Food and
    Development Policy
Moses T.K. Kairo • Florida A&M University
David Knopp • Emerging Markets Group (EMG)
Russ Kruska • International Livestock Research Institute
Andrew D.B. Leakey • University of Illinois
Karen Luz • World Wildlife Fund
Uford Madden • Florida A&M University
Pedro Marques • The World Bank
Harold J. McArthur • University of Hawaii at Manoa
A.J. McDonald • Cornell University
Patrick Meier • Tufts University
Douglas L. Murray • Colorado State University
Clare Narrod • International Food Policy Research Institute
James K. Newman • Iowa State University
Diane Osgood • Business for Social Responsibility
Jonathan Padgham • The World Bank
Harry Palmier • The World Bank
Philip Pardey • University of Minnesota
Ivette Perfecto • University of Michigan
Cameron Pittelkow • Independent
Carl E. Pray • Rutgers University
Elizabeth Ransom • University of Richmond
Laura T. Raynolds • Colorado State University
Peter Reich • University of Minnesota
Robin Reid • Colorado State University
Susan Riha • Cornell University
Claudia Ringler • International Food Policy Research Institute
Steven Rose • U.S. Environmental Protection Agency

Mark Rosegrant • International Food Policy Research Institute
Erika Rosenthal • Center for International Environmental Law
Michael Schechtman • U.S. Department of Agriculture
Sara Scherr • Ecoagriculture Partners
Jeremy Schwartzbord • Independent
Leonid Sharashkin • Independent
Matthew Spurlock • University of Massachusetts
Timothy Sulser • International Food Policy Research Institute
Steve Suppan • Institute for Agriculture and Trade Policy
Douglas L. Vincent • University of Hawaii at Manoa
Pai-Yei Whung • U.S. Department of Agriculture
David E. Williams • U.S. Department of Agriculture
Stan Wood • International Food Policy Research Institute
Angus Wright • California State University, Sacramento
Howard Yana Shapiro • MARS, Inc.
Stacey Young • U.S. Agency for International Development
Tingju Zhu • International Food Policy Research Institute

Uruguay

Gustavo Ferreira • Instituto Nacional de Investigación
    Agropecuaria (INIA), Tacuarembó
Luis Carlos Paolino • Technological Laboratory of Uruguay
    (LATU)
Lucía Pitalluga • University of the Republic

Uzbekistan

Sandjar Djalalov • Independent
Alisher A. Tashmatov • Ministry of Finance

Viet Nam

Duong Van Chin • The Cuulong Delta Rice Research Institute

Zambia

Charlotte Wonani • University of Zambia

Zimbabwe

Chiedza L. Muchopa • University of Zimbabwe
Lindela R. Ndlovu • National University of Science and
    Technology
Idah Sithole-Niang • University of Zimbabwe
Stephen Twomlow • International Crops Research Institute for
    the Semi-Arid Tropics

Annex C
Peer Reviewers

Argentina
Sandra Elizabeth Sharry • Universidad Nacional de La Plata

Australia
Government of Australia
Simon Hearn • Australian Centre for International Agricultural Research
Stuart Hill • University of Western Sydney
Tony Jansen • TerraCircle Inc.
Sarah Withers • Department of Foreign Affairs and Trade

Austria
Elfriede Fuhrmann • BMLFUW
Government of Austria

Benin
Shellemiah Keya • WARDA
Peter Neuenschwander • IITA

Brazil
Government of Brazil
Odo Primavesi • Embrapa Pecuaria Sudeste (Southeast Embrapa Cattle)
Francisco Reifschneider • Embrapa

Canada
David Cooper • Convention on Biological Diversity
Donald C. Cole • University of Toronto
Harriet • Friedman • University of Toronto
JoAnn Jaffe • University of Regina
Muffy Koch • Agbios
Iain C. MacGillivray • Canadian International Development Agency
Mary Stockdale • University of British Columbia, Okanagan

Denmark
Frands Dolberg • University of Arhus
Henrik Egelyng • Danish Institute for International Studies (DIIS)

Dominican Republic
Rufino Pérez-Brennan • ALIMENTEC S.A.

Egypt
Ayman Abou-Hadid • Agricultural Research Center

Finland
Riika Rajalahti • Ministry of Foreign Affairs

France
Louis Aumaitre • EAAP
Dominique Hervé • Institute for Development Research (IRD)
Jacques Loyat • Ministry of Agriculture
Michèle • Tixier-Boichard • Ministry of Higher Education and Research

Germany
Jan van Aken • Greenpeace International

India
Pradip Dey • Indian Council of Agricultural Research
Ramesh Chand • NCAP
C.P. Chandrasekhar • Jawaharlal Nehru University
Sudhir Kochhar • Indian Council of Agricultural Research
Aditya Misra • Project Directorate on Cattle
Suresh Pal • NCAP
C. Upendranadh • Institute for Human Development

Indonesia
Russell Dilts • Environmental Services Program

Iran
Farhad Saeidi Naeini • Iranian Research Institute of Plant Protection

Ireland
Government of Ireland
Sharon Murphy • Department of Agriculture, Fisheries and Food

Italy
Agriculture Department • FAO
Susan Braatz • FAO
Jorge Csirke • FAO
Forestry Department • FAO
Gender, Equity and Rural Employment Division of FAO
Yianna Lambrou • FAO
Shivaji Pandey • FAO
Teri Raney • FAO
Jeff Tschirley • FAO
Harry van der Wulp • FAO

Kenya
Christian Borgemeister • International Center for Insect Physiology and Ecology
Marcus Lee • United Nations Environment Programme
Evans Mwangi • University of Nairobi
Nalini Sharma • United Nations Environment Programme
Anna Stabrawa • United Nations Environment Programme

Madagascar
Xavier Rakotonjanahary • FOFIFA

Malaysia
Li Ching Lim • Third World Network

Nepal
Rajendra Shrestha • AFORDA

Netherlands
Judith Francis • Technical Centre for Agricultural and Rural
    Cooperation (CTA)
Juan Lopez Villar • Friends of the Earth International

New Zealand
A. Neil Macgregor • Journal of Organic Systems

Philippines
Leo Sebastian • Philippine Rice Research Institute

Poland
Ursula Soltysiak • AgroBio Test

Spain
Mario Giampietro • Universitat Autònoma de Barcelona
Marta Rivera-Ferre • Universitat Autònoma de Barcelona

Sweden
Ulf Herrström • Independent
Permilla Malmer • Swedish Biodiversity Center

Switzerland
David Duthie • United Nations Environment Programme

Tanzania
Jamidu Katima • University of Dar es Salaam

Tunisia
Rym Ben Zid • Independent

Uganda
Kevin Akoyi • Vredeseilanden
Henry Ssali • Kawanda Agricultural Research Institute

United Kingdom
Stephen Biggs • University of East Anglia
Janet Cotter • Greenpeace International, Exeter University
Stuart Coupe • Practical Action
Peter Craufurd • Reading University
Sue D'Arcy • Masterfoods UK
Department of Environment, Food and Rural Affairs
Department for International Development
Cathy Rozel Farnworth • Independent
Emma Hennessey • Defra
John Marsh • Independent
Clare Oxborrow • Friends of the Earth England, Wales and
    Northern Ireland
Helena Paul • EcoNexus
Pete Riley • GM Freeze
Jo Ripley • Independent
Geoff Tansey • Independent

Reyes Tirado • Greenpeace International
Stephanie Williamson • Pesticide Action Network, UK

United States
Miguel Altieri • University of California, Berkeley
Jock Anderson • The World Bank
Molly Anderson • Food Systems Integrity
Michael Arbuckle • The World Bank
Philip L. Bereano • University of Washington
David Bouldin • Cornell University
Lynn Brown • The World Bank
Rodney Brown • Brigham Young University
Glenn Carpenter • U.S. Department of Agriculture
Janet Carpenter • U.S. Department of Agriculture
Jean-Christophe Carret • The World Bank
Cheryl Christensen • U.S. Department of Agriculture
Nata Duvvury • International Center for Research on Women
Denis Ebodaghe • U.S. Department of Agriculture
Indira Ekanayake • The World Bank
Erick Fernandes • The World Bank
Steven Finch • U.S. Department of Agriculture
Mary-Ellen Foley • The World Bank
Lucia Fort • The World Bank
Christian Foster • U.S. Department of Agriculture
Bill Freese • Center for Food Safety
Government of the United States
Doug Gurian-Sherman • Union of Concerned Scientists
Michael Hansen • Consumers Union of US
Kenneth Hinga • U.S. Department of Agriculture
Gregory Jaffe • Center for Science in the Public Interest
Randy Johnson • U.S. Forest Service
Nadim Khouri • The World Bank
Jack Kloppenburg • University of Wisconsin
Masami Kojima • The World Bank
Anne Kuriakose • The World Bank
Saul Landau • California Polytechnic, Pomona
Jennifer Long • University of Illinois, Chicago
Karen Luz • World Wildlife Fund
William Martin • The World Bank
A.J. McDonald • Cornell University
Rekha Mehra • The World Bank
Douglas L. Murray • Colorado State University
Michael Naim • U.S. Department of Agriculture
John Nash • The World Bank
World Nieh • US Forest Service
Jon Padgham • World Bank
Mikko Paunio • The World Bank
Eija Pehu • The World Bank
Carl Pray • Rutgers University
Margaret Reeves • Pesticide Action Network North America
Peter Riggs • Forum on Democracy & Trade
Naomi Roht-Arriaza • University of California Hastings College
    of Law
Phrang Roy • The Christensen Fund
Marc Safley • U.S. Department of Agriculture
Michael Schechtman • U.S. Department of Agriculture
Sara Scherr • Ecoagriculture Partners
Seth Shames • Ecoagriculture Partners
Doreen Stabinsky • College of the Atlantic
Lorann Stallones • Colorado State University
Gwendolyn H. Urey • California Polytechnic, Pomona

Evert Van der Sluis • South Dakota State University
David Winickoff • University of California, Berkeley
Angus Wright • California State University, Sacramento
Stacey Young • U.S. Agency International Development

Annex D
Secretariat and Cosponsor Focal Points

Secretariat

World Bank
Marianne Cabraal, Leonila Castillo, Jodi Horton, Betsi Isay, Pekka Jamsen, Pedro Marques, Beverly McIntyre, Wubi Mekonnen, June Remy

UNEP
Marcus Lee, Nalini Sharma, Anna Stabrawa

UNESCO
Guillen Calvo

With special thanks to the Publications team: Audrey Ringler (logo design), Pedro Marques (proofing and graphics), Ketill Berger and Eric Fuller (graphic design)

Regional Institutes

Sub-Saharan Africa – African Centre for Technology Studies (ACTS)
Ronald Ajengo, Elvin Nyukuri, Judi Wakhungu

Central and West Asia and North Africa – International Center for Agricultural Research in the Dry Areas (ICARDA)
Mustapha Guellouz, Lamis Makhoul, Caroline Msrieh-Seropian, Ahmed Sidahmed, Cathy Farnworth

Latin America and the Caribbean – Inter-American Institute for Cooperation on Agriculture (IICA)
Enrique Alarcon, Jorge Ardila Vásquez, Viviana Chacon, Johana Rodríguez, Gustavo Sain

East and South Asia and the Pacific – WorldFish Center
Karen Khoo, Siew Hua Koh, Li Ping Ng, Jamie Oliver, Prem Chandran Venugopalan

Cosponsor Focal Points

GEF	Mark Zimsky
UNDP	Philip Dobie
UNEP	Ivar Baste
UNESCO	Salvatore Arico, Walter Erdelen
WHO	Jorgen Schlundt
World Bank	Mark Cackler, Kevin Cleaver, Eija Pehu, Juergen Voegele

Annex E
Steering Committee for Consultative Process and Advisory Bureau for Assessment

Steering Committee

The Steering Committee was established to oversee the consultative process and recommend whether an international assessment was needed, and if so, what was the goal, the scope, the expected outputs and outcomes, governance and management structure, location of the Secretariat and funding strategy.

Co-chairs

Louise Fresco, Assistant Director General for Agriculture, FAO
Seyfu Ketema, Executive Secretary, Association for Strengthening Agricultural Research in East and Central Africa (ASARECA)
Claudia Martinez Zuleta, Former Deputy Minister of the Environment, Colombia
Rita Sharma, Principal Secretary and Rural Infrastructure Commissioner, Government of Uttar Pradesh, India
Robert T. Watson, Chief Scientist, The World Bank

Nongovernmental Organizations

Benny Haerlin, Advisor, Greenpeace International
Marcia Ishii-Eiteman, Senior Scientist, Pesticide Action Network North America Regional Center (PANNA)
Monica Kapiriri, Regional Program Officer for NGO Enhancement and Rural Development, Aga Khan
Raymond C. Offenheiser, President, Oxfam America
Daniel Rodriguez, International Technology Development Group (ITDG), Latin America Regional Office, Peru

UN Bodies

Ivar Baste, Chief, Environment Assessment Branch, UN Environment Programme
Wim van Eck, Senior Advisor, Sustainable Development and Healthy Environments, World Health Organization
Joke Waller-Hunter, Executive Secretary, UN Framework Convention on Climate Change
Hamdallah Zedan, Executive Secretary, UN Convention on Biological Diversity

At-large Scientists

Adrienne Clarke, Laureate Professor, School of Botany, University of Melbourne, Australia
Denis Lucey, Professor of Food Economics, Dept. of Food Business & Development, University College Cork, Ireland, and Vice-President NATURA
Vo-tong Xuan, Rector, Angiang University, Vietnam

Private Sector

Momtaz Faruki Chowdhury, Director, Agribusiness Center for Competitiveness and Enterprise Development, Bangladesh

Sam Dryden, Managing Director, Emergent Genetics
David Evans, Former Head of Research and Technology, Syngenta International
Steve Parry, Sustainable Agriculture Research and Development Program Leader, Unilever
Mumeka M. Wright, Director, Bimzi Ltd., Zambia

Consumer Groups

Michael Hansen, Consumers International
Greg Jaffe, Director, Biotechnology Project, Center for Science in the Public Interest
Samuel Ochieng, Chief Executive, Consumer Information Network

Producer Groups

Mercy Karanja, Chief Executive Officer, Kenya National Farmers' Union
Prabha Mahale, World Board, International Federation Organic Agriculture Movements (IFOAM)
Tsakani Ngomane, Director Agricultural Extension Services, Department of Agriculture, Limpopo Province, Republic of South Africa
Armando Paredes, Presidente, Consejo Nacional Agropecuario (CNA)

Scientific Organizations

Jorge Ardila Vásquez, Director Area of Technology and Innovation, Inter-American Institute for Cooperation on Agriculture (IICA)
Samuel Bruce-Oliver, NARS Senior Fellow, Global Forum for Agricultural Research Secretariat
Adel El-Beltagy, Chair, Center Directors Committee, Consultative Group on International Agricultural Research (CGIAR)
Carl Greenidge, Director, Center for Rural and Technical Cooperation, Netherlands
Mohamed Hassan, Executive Director, Third World Academy of Sciences (TWAS)
Mark Holderness, Head Crop and Pest Management, CAB International
Charlotte Johnson-Welch, Public Health and Gender Specialist and Nata Duvvury, Director Social Conflict and Transformation Team, International Center for Research on Women (ICRW)
Thomas Rosswall, Executive Director, International Council for Science (ICSU)
Judi Wakhungu, Executive Director, African Center for Technology Studies

Governments

Australia: Peter Core, Director, Australian Centre for International Agricultural Research

China: Keming Qian, Director General Inst. Agricultural Economics, Dept. of International Cooperation, Chinese Academy of Agricultural Science

Finland: Tiina Huvio, Senior Advisor, Agriculture and Rural Development, Ministry of Foreign Affairs

France: Alain Derevier, Senior Advisor, Research for Sustainable Development, Ministry of Foreign Affairs

Germany: Hans-Jochen de Haas, Head, Agricultural and Rural Development, Federal Ministry of Economic Cooperation and Development (BMZ)

Hungary: Zoltan Bedo, Director, Agricultural Research Institute, Hungarian Academy of Sciences

Ireland: Aidan O'Driscoll, Assistant Secretary General, Department of Agriculture and Food

Morocco: Hamid Narjisse, Director General, INRA

Russia: Eugenia Serova, Head, Agrarian Policy Division, Institute for Economy in Transition

Uganda: Grace Akello, Minister of State for Northern Uganda Rehabilitation

United Kingdom Paul Spray, Head of Research, DFID

United States: Rodney Brown, Deputy Under Secretary of Agriculture and Hans Klemm, Director of the Office of Agriculture, Biotechnology and Textile Trade Affairs, Department of State

Foundations and Unions

Susan Sechler, Senior Advisor on Biotechnology Policy, Rockefeller Foundation

Achim Steiner, Director General, The World Conservation Union (IUCN)

Eugene Terry, Director, African Agricultural Technology Foundation

Advisory Bureau

Non-government Representatives

Consumer Groups
Jaime Delgado • Asociación Peruana de Consumidores y Usuarios
Greg Jaffe • Center for Science in the Public Interest
Catherine Rutivi • Consumers International
Indrani Thuraisingham • Southeast Asia Council for Food Security and Trade
Jose Vargas Niello • Consumers International Chile

International organizations
Nata Duvvury • International Center for Research on Women
Emile Frison • CGIAR
Mohamed Hassan • Third World Academy of Sciences
Mark Holderness • GFAR
Jeffrey McNeely • World Conservation Union (IUCN)
Dennis Rangi • CAB International
John Stewart • International Council of Science (ICSU)

NGOs
Kevin Akoyi • Vredeseilanden
Hedia Baccar • Association pour la Protection de l'Environment de Kairouan
Benedikt Haerlin • Greenpeace International
Juan Lopez • Friends of the Earth International
Khadouja Mellouli • Women for Sustainable Development
Patrick Mulvaney • Practical Action
Romeo Quihano • Pesticide Action Network
Maryam Rahmaniam • CENESTA
Daniel Rodriguez • International Technology Development Group

Private Sector
Momtaz Chowdhury • Agrobased Technology and Industry Development
Giselle L. D'Almeida • Interface
Eva Maria Erisgen • BASF
Armando Paredes • Consejo Nacional Agropecuario
Steve Parry • Unilever
Harry Swaine • Syngenta (resigned)

Producer Groups
Shoaib Aziz • Sustainable Agriculture Action Group of Pakistan
Philip Kiriro • East African Farmers Federation
Kristie Knoll • Knoll Farms
Prabha Mahale • International Federation of Organic Agriculture Movements
Anita Morales • Apit Tako
Nizam Selim • Pioneer Hatchery

Government Representatives

Central and West Asia and North Africa
Egypt • Ahlam Al Naggar
Iran • Hossein Askari
Kyrgyz Republic • Djamin Akimaliev
Saudi Arabia • Abdu Al Assiri, Taqi Elldeen Adar, Khalid Al Ghamedi
Turkey • Yalcin Kaya, Mesut Keser

East/South Asia/Pacific
Australia • Simon Hearn
China • Puyun Yang
India • PK Joshi
Japan • Ryuko Inoue
Philippines • William Medrano

Latin America and Caribbean
Brazil • Sebastiao Barbosa, Alexandre Cardoso, Paulo Roberto Galerani, Rubens Nodari
Dominican Republic • Rafael Perez Duvergé
Honduras • Arturo Galo, Roberto Villeda Toledo
Uruguay • Mario Allegri

North America and Europe
Austria • Hedwig Woegerbauer
Canada • Iain MacGillivray
Finland • Marja-Liisa Tapio-Bistrom
France • Michel Dodet
Ireland • Aidan O'Driscoll, Tony Smith
Russia • Eugenia Serova, Sergey Alexanian
United Kingdom • Jim Harvey, David Howlett, John Barret
United States • Christian Foster

Sub-Saharan Africa
Benin • Jean Claude Codjia
Gambia • Sulayman Trawally
Kenya • Evans Mwangi
Mozambique • Alsácia Atanásio, Júlio Mchola
Namibia • Gillian Maggs-Kölling
Senegal • Ibrahim Diouck

About Island Press

Since 1984, the nonprofit Island Press has been stimulating, shaping, and communicating the ideas that are essential for solving environmental problems worldwide. With more than 800 titles in print and some 40 new releases each year, we are the nation's leading publisher on environmental issues. We identify innovative thinkers and emerging trends in the environmental field. We work with world-renowned experts and authors to develop cross-disciplinary solutions to environmental challenges.

Island Press designs and implements coordinated book publication campaigns in order to communicate our critical messages in print, in person, and online using the latest technologies, programs, and the media. Our goal: to reach targeted audiences—scientists, policymakers, environmental advocates, the media, and concerned citizens—who can and will take action to protect the plants and animals that enrich our world, the ecosystems we need to survive, the water we drink, and the air we breathe.

Island Press gratefully acknowledges the support of its work by the Agua Fund, Inc., Annenberg Foundation, The Christensen Fund, The Nathan Cummings Foundation, The Geraldine R. Dodge Foundation, Doris Duke Charitable Foundation, The Educational Foundation of America, Betsy and Jesse Fink Foundation, The William and Flora Hewlett Foundation, The Kendeda Fund, The Forrest and Frances Lattner Foundation, The Andrew W. Mellon Foundation, The Curtis and Edith Munson Foundation, Oak Foundation, The Overbrook Foundation, the David and Lucile Packard Foundation, The Summit Fund of Washington, Trust for Architectural Easements, Wallace Global Fund, The Winslow Foundation, and other generous donors.

The opinions expressed in this book are those of the author(s) and do not necessarily reflect the views of our donors.